REAL ANALYSIS

REAL ANALYSIS

Shanzhen Lu
Kunyang Wang
Beijing Normal University, China

Translators

Moyan Qin, Jiawei Tan, Jiahui Wang, Yuru Li,
Xiaoting Qiu, Shenglong Lin, Yuan Zhao

北京师范大学出版集团
BEIJING NORMAL UNIVERSITY PUBLISHING GROUP
北京师范大学出版社

World Scientific

Published by

World Scientific Publishing Co. Pte. Ltd.
5 Toh Tuck Link, Singapore 596224
USA office: 27 Warren Street, Suite 401-402, Hackensack, NJ 07601
UK office: 57 Shelton Street, Covent Garden, London WC2H 9HE

Library of Congress Control Number: 2024059312

British Library Cataloguing-in-Publication Data
A catalogue record for this book is available from the British Library.

实分析
Originally published in Chinese by Beijing Normal University Press (Group) Co., Ltd.
Copyright © Beijing Normal University Press (Group) Co., Ltd., 2006

REAL ANALYSIS

ISBN 978-981-12-9738-0 (hardcover)
ISBN 978-981-12-9739-7 (ebook for institutions)
ISBN 978-981-12-9740-3 (ebook for individuals)

For any available supplementary material, please visit
https://www.worldscientific.com/worldscibooks/10.1142/13965#t=suppl

Desk Editors: Nambirajan Karuppiah/Angeline Husni

Typeset by Stallion Press
Email: enquiries@stallionpress.com

About the Authors

Shanzhen Lu, born in November 1939, is a well-known mathematician and educator. He graduated from East China Normal University in July 1961 and began teaching in the Department of Mathematics of Beijing Normal University in China in September of the same year. From 1980 to 1982, he was a visiting scholar at Washington University in St. Louis, where he was appointed as a Professor of Mathematics. He was promoted from Lecturer to Professor at Beijing Normal University in May 1983 and approved by the Academic Degrees Committee of the State Council of the People's Republic of China as a Doctoral Supervisor in January 1984. He has successively served as the editorial board member of *Science China, Chinese Science Bulletin, Acta Mathematica Sinica, Frontiers of Mathematics in China* and a number of foreign mathematical journals, Deputy Editor of *Advances in Mathematics*, and Editor-in-Chief of *Pure Mathematics*. In addition, he has served as the Chairman of major harmonic analysis conferences at home and abroad.

Professor Lu has long been engaged in the research of harmonic analysis and has made important contributions to the development of this field and many praised research results by international peer experts during his academic career of more than 50 years. Among them, the most prominent work is his research work on the Bochner–Riesz means at critical index, which has a wide impact on this field. Professor Lu and his collaborators created and made full use of the block space theory to completely solve Fefferman's conjecture about the almost everywhere convergence of the Bochner–Riesz means at critical index, which was cited by famous mathematicians, such as Stein, Meyer and others. Besides, Professor Lu proved the localized convergence theorem for the q-order strong summation of Bochner–Riesz means at critical index, which Fefferman commented as follows: This problem has not been solved for 36 years since the works of Bochner and Chandrasekharan in 1948. Tao's articles on the almost convergence of the Bochner–Riesz means also highlights Professor Lu's research work about it. In addition, Professor Lu and his students studied oscillatory singular integrals with rough kernels and gave a discriminant criterion about their boundedness in L^p spaces.

Professor Lu is an excellent mathematics educator who has trained many outstanding students, such as Professor Heping Liu of Peking University in China, Professor Dachun Yang and Professor Yong Ding of Beijing Normal University, Professor Dunyan Yan of the University of Chinese Academy of Sciences, and Professor Xiaochun Li of the University of Illinois at Urbana-Champaign, who have made important achievements in their respective fields.

Kunyang Wang, national model of teaching in China, graduated from the Department of Mathematical Mechanics of Peking University in China in 1966. In 1978, he was admitted to Beijing Normal University, where he completed his master's and doctoral studies, under the tutelage of Professor Yongsheng Sun. He went to Germany, Australia, Canada and other countries as a Senior Visiting Scholar or Visiting Professor many times. Professor Wang has served as a member of the Beijing Municipal Committee of the Chinese People's Political Consultative Conference, a member of the Mathematics Subcommittee of the Higher Education Mathematics and Statistics Teaching Committee of the Ministry of Education of China, the Director of the Education Work Committee of the Chinese Mathematical Society, the editorial board member of journals such as *Advances in Mathematics* (China) and *Journal of Mathematical Research and Exposition* (China), and has participated in the reform of national teaching projects for many times.

Professor Wang has published more than 60 academic papers, two academic monographs and three textbooks. The Chinese textbooks he compiled, such as *Real Analysis, Concise Mathematical Analysis, Special Research on Mathematical Analysis, Lecture Notes on Mathematical Analysis, Mathematical Analysis*, and *Orthogonal Series*, have been deeply preferred by teachers and students. He has also successively published eight teaching reform papers, hosted and completed the key project of teaching reform of the Teacher Normal Department of the Ministry of Education of China, the famous course project created by the National Science Base in China, and the excellent course project of Mathematical Analysis. He was rated as an advanced worker by the Ministry of Education and the National Natural Science Foundation of China and won the Beijing Famous Teacher Award.

Contents

Chapter 1

Abstract Measure and Integral

In the undergraduate course of "theory of functions of real variable", we have learned the Lebesgue integral theory on $\mathbb{R}^n$. In this theory, we first establish the concept of **outer measure** on $\mathbb{R}^n$. The basis of the concept of outer measure is that the "volume" of a cube in $\mathbb{R}^n$ is equal to the nth power of its side length. Every subset of $\mathbb{R}^n$ has the outer measure. The outer measure of an empty set is zero, and the outer measure is a nonnegative generalized real-valued set function with the property of **countable subadditivity**. After the outer measure is established, the concept of measurable set can be derived. A set satisfying the Carathéodory condition is measurable. For a measurable set, its outer measure is called measure. Empty set and $\mathbb{R}^n$ are measurable sets, and their measures are zero and ∞, respectively. The family of measurable sets is closed under difference and countable unions. With the concept of measure, measurable functions and integrals can be defined and studied in depth by the established integral theory. References [12, 14] for more information.

The purpose of this chapter is to abstract the content of the above Lebesgue integral theory, establishing the concept of measure on the general set. Thus, the theory is extended to the general abstract set, making it applicable to a wider range of objects and get a wider range of applications.

1.1   Measure

Definition 1.1.1. Given a set X, a nonempty collection $\mathscr{R}$ of subsets of X is called a σ-ring on X provided

(1) $A, B \in \mathscr{R} \Rightarrow A\backslash B \in \mathscr{R}$;
(2) $A_n \in \mathscr{R}, n \in \mathbb{N} \Rightarrow \bigcup_{n=1}^{\infty} A_n \in \mathscr{R}$.

We say $\mathscr{A}$ is a σ-algebra on X if $\mathscr{A}$ is a σ-ring such that $X \in \mathscr{A}$.

It is obvious that a σ-ring must contain the empty set $\emptyset$, and

$$\bigcap_{n=1}^{\infty} A_n = \bigcup_{n=1}^{\infty} A_n \backslash \bigcup_{n=1}^{\infty} \left(\bigcup_{k=1}^{\infty} A_k \backslash A_n \right),$$

given by countable intersections and unions, and differences of sets. Therefore, $\mathscr{R}$ is closed under countable intersections. However, $\mathscr{R}$ does not have to be closed under complements because $\mathscr{R}$ does not necessarily contain X, and the σ-algebra $\mathscr{A}$ must be closed under complements.

Definition 1.1.2. If $\mathscr{A}$ is a σ-algebra on a set X, then $(X, \mathscr{A})$ is called a measurable space.

As done in the course, "theory of functions of real variable",

$$\overline{\mathbb{R}} = \mathbb{R} \cup \{-\infty, \infty\}$$

is specified as the generalized real number system, which has two more symbolic elements, negative infinity $-\infty$ and positive infinity ∞, than the real number system $\mathbb{R}$. $-\infty$ and ∞ are not numbers, but as a rule, they can be compared with real numbers and participate in appropriate arithmetic operations. References [5,6] for more information. Here, we only point out that the result of multiplying zero by infinity ($-\infty$ or ∞) is zero.

Definition 1.1.3. Suppose $\mathscr{R}$ is a σ-ring on X and φ is an injective mapping (function) from $\mathscr{R}$ to $\overline{\mathbb{R}}$. φ is called a signed measure on $\mathscr{R}$ provided $\varphi(\emptyset) = 0$ and φ is countably additive, i.e. $\varphi(\bigcup_{n=1}^{\infty} A_n) = \sum_{n=1}^{\infty} \varphi(A_n)$ if $A_n \in \mathscr{R}, n \in \mathbb{N}$ and $A_m \cap A_n = \emptyset$ (for all $m, n \in \mathbb{N}, m \neq n$). A signed measure that does not take negative values is called a measure. A space $(X, \mathscr{A}, \mu)$ is called a measure space provided $(X, \mathscr{A})$ is a measurable space and μ is a measure on $\mathscr{A}$. Let $(X, \mathscr{A}, \mu)$ be a measure space, elements of $\mathscr{A}$ are called μ-measurable sets or measurable sets for short. A measure space $(X, \mathscr{A}, \mu)$ is called complete (or that μ is complete) provided it makes all subsets of sets of measure zero measurable.

Remark. For a signed measure φ, it is impossible to take both values ∞ and $-\infty$ at the same time. Otherwise, let $\varphi(A) = \infty, \varphi(B) = -\infty$, then $\varphi(A \cup B) + \varphi(A \cap B) \neq \varphi(A) + \varphi(B)$.

Definition 1.1.4. Suppose $(X, \mathscr{A})$ is a measurable space. A signed measure φ on $\mathscr{A}$ is called finite if it takes only finite values and σ-finite if $X = \cup_{k \in \mathbb{N}} E_k$ where $E_k \in \mathscr{A}$ and $\varphi(E_k) \in \mathbb{R}$ for all $k \in \mathbb{N}$.

Example 1.1.1. Let X be a set. For $A \subset X$, define

$$\nu(A) = \begin{cases} \text{card}(A) & \text{if } A \text{ is finite,} \\ \infty & \text{if } A \text{ is infinite,} \end{cases}$$

where $\text{card}(A)$ represents the cardinal of the set A and 2^X represents the set consisting of all subsets of X. Then ν is a measure on 2^X, which is referred to as counting measure.

Example 1.1.2. Let X be a set. For $A \subset X$, define

$$\mu(A) = \begin{cases} 0 & \text{if } A = \emptyset, \\ \infty & \text{if } A \neq \emptyset, \end{cases}$$

then μ is a measure on 2^X. In fact, μ is a degenerate measure.

The basic properties of measure are discussed in the following.

Theorem 1.1.1. *Let μ be a measure on σ-ring $\mathscr{R}$. If $A, B \in \mathscr{R}$ with $A \subset B$, then $\mu(A) \leq \mu(B)$.*

Proof. $\mu(B) = \mu(A) + \mu(B \backslash A) \geq \mu(A)$. $\qquad \square$

Theorem 1.1.2. *Let μ be a measure on σ-ring $\mathscr{R}$ and $\{A_n\}_{n=1}^{\infty}$ be a monotonic increasing sequence in $\mathscr{R}$. Then*

$$\mu\left(\lim_{n \to \infty} A_n\right) = \lim_{n \to \infty} \mu(A_n).$$

Proof.   Let $A_0 = \emptyset, B_n = A_n \backslash A_{n-1}, n \in \mathbb{N}$, then

$$\lim_{n \to \infty} A_n = \bigcup_{n=1}^{\infty} B_n,$$

$B_m \cap B_n = \emptyset$ when $m, n \in \mathbb{N}$ with $m \neq n$, and $B_n \in \mathscr{R}$. By the countable additivity of measures,

$$\mu\left(\lim_{n \to \infty} A_n\right) = \sum_{n=1}^{\infty} \mu(B_n) = \lim_{n \to \infty} \sum_{k=1}^{n} \mu(B_k) = \lim_{n \to \infty} \mu(A_n).$$

$\square$

Theorem 1.1.3. *Let μ be a measure on σ-ring $\mathscr{R}$ and $\{A_n\}_{n=1}^{\infty}$ be a monotonically decreasing sequence in $\mathscr{R}$ with $\mu(A_1) < \infty$, then*

$$\mu\left(\lim_{n \to \infty} A_n\right) = \lim_{n \to \infty} \mu(A_n).$$

Proof.   Let $B_n = A_1 \backslash A_n$, then

$$\mu\left(\lim_{n \to \infty} B_n\right) = \lim_{n \to \infty} \mu(B_n) = \lim_{n \to \infty} \left(\mu(A_1) \backslash \mu(A_n)\right)$$

$$= \mu(A_1) - \lim_{n \to \infty} \mu(A_n)$$

by Theorem 1.1.2. Note that

$$\lim_{n \to \infty} B_n = \bigcup_{n=1}^{\infty} (A_1 \backslash A_n) = A_1 \backslash \bigcap_{n=1}^{\infty} A_n = A_1 \backslash \lim_{n \to \infty} A_n.$$

Thus,

$$\mu(A_1) - \mu\left(\lim_{n \to \infty} A_n\right) = \mu(A_1) - \lim_{n \to \infty} \mu(A_n),$$

that is,

$$\mu\left(\lim_{n \to \infty} A_n\right) = \lim_{n \to \infty} \mu(A_n).$$

$\square$

Remark. The condition, $\mu(A_1) < \infty$, in Theorem 1.1.3 is indispensable. For example, for a Lebesgue measure m on $\mathbb{R}$ and a sequence of sets $\{(n, \infty)\}_{n=1}^{\infty}$, we have

$$m\left(\lim_{n \to \infty} (n, \infty)\right) = 0 < \lim_{n \to \infty} m((n, \infty)) = \infty.$$

Theorem 1.1.4. *Let μ be a measure on a σ-ring $\mathscr{R}$ and $A_n \in \mathscr{R}$, $n \in \mathbb{N}$. Then*

$$\mu\left(\varliminf_{n\to\infty} A_n\right) \le \varliminf_{n\to\infty} \mu(A_n).$$

Further, if $\mu(\cup_{n=1}^{\infty} A_n) < \infty$, then

$$\mu\left(\varlimsup_{n\to\infty} A_n\right) \ge \varlimsup_{n\to\infty} \mu(A_n).$$

Proof. Note that

$$\varliminf_{n\to\infty} A_n = \bigcup_{n=1}^{\infty}\bigcap_{k=n}^{\infty} A_k = \lim_{n\to\infty}\left(\bigcap_{k=n}^{\infty} A_k\right),$$

$$\varlimsup_{n\to\infty} A_n = \bigcap_{n=1}^{\infty}\bigcup_{k=n}^{\infty} A_k = \lim_{n\to\infty}\left(\bigcup_{k=n}^{\infty} A_k\right),$$

and

$$\bigcup_{k=n}^{\infty} A_k \supset A_n, \quad \bigcap_{k=n}^{\infty} A_k \subset A_n,$$

the desired conclusions can be immediately derived by Theorems 1.1.1–1.1.3. $\qquad\square$

Exercise 1.1

1. Let Ω be a countable set and $\mathscr{F}$ be a collection of all finite subsets of Ω and their complements. Prove that $\mathscr{F}$ is not a σ-algebra but closed under finite unions, intersections, differences and complements (such a nonempty collection is called an algebra).
2. Let μ be a nonnegative and finite additive function with respect to sets defined on a σ-algebra $\mathscr{A}$, i.e. $\mu(A \cup B) = \mu(A) + \mu(B)$ if $A, B \in \mathscr{A}$ with $A \cap B = \emptyset$. Prove that if $\{A_n\}_{n=1}^{\infty}$ is a pairwise disjoint collection of $\mathscr{A}$, then

$$\mu\left(\bigcup_{n=1}^{\infty} A_n\right) \ge \sum_{n=1}^{\infty} \mu(A_n).$$

Give an example of making the unequal sign true.

3. Let $(X, \mathscr{A}, \mu)$ be a finite measure space and $E_1, E_2 \in \mathscr{A}$. The sets E_1 and E_2 are treated as the same set if $\mu(E_1 \triangle E_2) = 0$, where $E_1 \triangle E_2 := (E_1 \backslash E_2) \cup (E_2 \backslash E_1)$. Stipulate $d(E_1, E_2) = \mu(E_1 \triangle E_2)$ as the distance between E_1 and E_2 which are in $\mathscr{A}$. Show that $(\mathscr{A}, d)$ is a complete metric space.

1.2   Measurable Function and Integration

Definition 1.2.1. Let $\mathscr{A}$ be a σ-algebra on a set X and $E \in \mathscr{A}$. A generalized real-valued function f defined on E, i.e. f is an injective mapping from E to $\overline{\mathbb{R}}$, is called a $\mathscr{A}$-measurable function, or just a measurable function if the sets $\{x \in E : f(x) > a\} \in \mathscr{A}$ for all $a \in \mathbb{R}$.

Theorem 1.2.1. *Let $\mathscr{A}$ be a σ-algebra on a set X and $E \in \mathscr{A}$, then the following conditions hold:*

(1) *All $\mathscr{A}$-measurable functions, which take finite values only, form a real linear space.*

(2) *If f is a measurable function on E, then*

$$f^+ := \max(f, 0) := f \vee 0,$$

$$f^- := \min(f, 0) := -(f \wedge 0),$$

$$|f|, \ |f|^p (p > 0)$$

are all measurable functions on E, where "$:=$" means definition, i.e. the content to the right of it is the definition of the symbol on its left, and $\frac{1}{f}$ is also a measurable function if $f(x) \neq 0$ for any $x \in E$.

(3) *If $\{f_n\}_{n \in \mathbb{N}}$ are measurable functions on E, then*

$$\sup\{f_n : n \in \mathbb{N}\}, \quad \inf\{f_n : n \in \mathbb{N}\}, \quad \varlimsup_{n \to \infty} f_n, \ \varliminf_{n \to \infty} f_n$$

are all measurable functions.

(4) *If f and g are measurable on E, then fg is measurable on E.*

The proof of Theorem 1.2.1 is so easy that it is omitted.

Denote the characteristic function of the set E by χ_E.

Definition 1.2.2. Let $\mathscr{A}$ be a σ-algebra on the set X, $E_1, \ldots, E_n$ be pairwise disjoint sets such that $\cup_{i=1}^n E_i = X$ and $a_1, \ldots, a_n$ be real numbers that are different from each other, then the function

$$\varphi(x) = \sum_{i=1}^{n} a_i \chi_{E_i}(x), \quad x \in X \qquad (*)$$

is called a simple function, i.e. a simple function on X is a finite linear combination of real-valued measurable ($\mathscr{A}$-measurable) functions. The equation $(*)$ is called the standard representation of the simple function φ.

Theorem 1.2.2. *Suppose f is a nonnegative measurable function on the measurable set E. Then there exists a sequence $\{f_n\}_{n \in \mathbb{N}}$ of nonnegative simple functions such that $f_n \nearrow f$ on E.*

The proof of Theorem 1.2.2 is the same as that learned in theory of functions of real variable, see [1, 9] for details, so it is omitted.

Theorem 1.2.3 (Egorov's theorem). *Let $(X, \mathscr{A}, \mu)$ be a measure space, $E \in \mathscr{A}$ and $\mu(E) < \infty$. Suppose $f_n, n \in \mathbb{N}$ are measurable and finite μ-a.e. on E, i.e. every f_n is finite almost everywhere on E except for a subset of μ-measure zero. If $\{f_n\}$ converges μ-a.e. on E, then for any $\varepsilon > 0$ there exists a set $A \subset E$ such that $\mu(E \backslash A) < \varepsilon$ and $\{f_n\}$ converges uniformly on A.*

The proof of Theorem 1.2.3 is the same as that learned in theory of functions of real variable.

Definition 1.2.3. Let $(X, \mathscr{A}, \mu)$ be a measure space, f and $f_n, n \in \mathbb{N}$ be $\mathscr{A}$-measurable functions on X. The function f_n is said to converge to f in measure μ, denoted by $f_n \xrightarrow{\mu} f$, provided for any $\delta > 0$,

$$\lim_{n \to \infty} \mu(\{x \in X : |f(x) - f_n(x)| \geq \delta\}) = 0.$$

Theorem 1.2.4 (Riesz's theorem). *If $f_n \xrightarrow{\mu} f$, then there exists a subsequence $\{f_{n_k}\}_{k=1}^{\infty}$ converging to f μ-a.e.*

Remark. The terminology, "μ-a.e." or "a.e." for short, means almost everywhere in measure μ. To say that a proposition related to an independent variable holds μ-a.e. or a.e. means that the proposition holds for every point except for a set of μ-measure 0. The proof of Theorem 1.2.4 has been shown in theory of functions of real variable. In this chapter, the Luzin theorem of Theory of Functions of

Real Variable will not be discussed since measure spaces do not have
a topological structure. But in Chapter 2, we will see corresponding generalizations of the Luzin theorem in appropriate occasions
when discussing measures over topological spaces. The following is
the theory of Lebesgue integration over general measure spaces.

Definition 1.2.4. Let $(X, \mathscr{A}, \mu)$ be a measure space.

(1) If φ is a nonnegative simple function with the standard
representation

$$\varphi(x) = \sum_{i=1}^{n} a_i \chi_{E_i}, \qquad (*)$$

then $\sum_{i=1}^{n} a_i \mu(E_i)$ is called the integral of φ on X, denoted by
$\int_X \varphi d\mu$.

(2) If f is a nonnegative measurable function on X, then

$$\sup \left\{ \int_X \varphi d\mu : \varphi \text{ is a simple function, } 0 \leq \varphi \leq f \right\}$$

is called the integral of f on X, denoted by $\int_X f d\mu$.

(3) Let f be a measurable function on X. Define

$$\int_X f d\mu = \int_X f^+ d\mu - \int_X f^- d\mu$$

as the integral of f on X provided $\int_X f^+ d\mu$ and $\int_X f^- d\mu$ are not
equal to ∞ at the same time. The measurable function f is said
to be integrable on X if the value of $\int_X f d\mu$ is a real number.
Denote the set of all integrable functions by $L(X, \mathscr{A}, \mu)$.

It is clear that Definitions 1.1.1–1.1.3 have abstracted the essence
of the Lebesgue measure on $\mathbb{R}^n$ taught in theory of functions of
real variable, see [2, 7]. The general notion of measure is established
with this essential property as a definition. On this base,
Definition 1.2.4 is identical to the definition of the Lebesgue integral given in theory of functions of real variable. Thus, the whole
theory of Lebesgue integral unfolds afterwards in exactly the same
way as the usual integral theory on $\mathbb{R}^n$. Henceforth, we will refer to
the notion of measure on $\mathbb{R}^n$ based on the volume of a cube, the

nth power of the side length of the cube, and the corresponding theory, which are described in theory of functions of real variable, as the usual Lebesgue measure and the usual Lebesgue integral. All the conclusions of the usual Lebesgue theory can be transferred here, as long as they do not involve the topological structure of $\mathbb{R}^n$ and the specific way in which the measures on $\mathbb{R}^n$ are taken, since there is no topological structure in the measure space here, which are listed in the following. The content involved in these conclusions is not considered within the scope of this book, so the proof is omitted, see theory of functions of real variable for details.

Theorem 1.2.5. *Let $(X, \mathscr{A}, \mu)$ be a measure space, and f and g be measurable functions on X. If $\int_X f d\mu$ and $\int_X g d\mu$ exist, then*

(1) $f \le g \Rightarrow \int_X f d\mu \le \int_X g d\mu$;
(2) $\mu(X) = 0 \Rightarrow \int_X f d\mu = 0$;
(3) $f = 0$ μ-a.e. $\Leftrightarrow \int_X |f| d\mu = 0$.

Theorem 1.2.6. *The space $L(X, \mathscr{A}, \mu)$ is a real linear space.*

Theorem 1.2.7 (Levi's theorem). *Let $k \in \mathbb{N}$, f_k be measurable and $f_k \to f(k \to \infty)$, then*

(1) $L(X, \mathscr{A}, \mu) \ni \varphi \le f_k \nearrow f \Rightarrow \int_X f_k d\mu \longrightarrow \int_X f d\mu$;
(2) $L(X, \mathscr{A}, \mu) \ni \varphi \ge f_k \searrow f \Rightarrow \int_X f_k d\mu \longrightarrow \int_X f d\mu$.

Theorem 1.2.8 (Fatou's theorem). *Let $k \in \mathbb{N}$ and f_k be measurable, then*

(1) $L(X, \mathscr{A}, \mu) \ni \varphi \le f_k \Rightarrow \int_X \underline{\lim}_{k \to \infty} f_k d\mu \le \underline{\lim}_{k \to \infty} \int_X f_k d\mu$;
(2) $L(X, \mathscr{A}, \mu) \ni \varphi \ge f_k \Rightarrow \int_X \overline{\lim}_{k \to \infty} f_k d\mu \ge \overline{\lim}_{k \to \infty} \int_X f_k d\mu$.

Theorem 1.2.9 (Lebesgue's theorem). *Let $k \in \mathbb{N}$, f_k be $\mathscr{A}$-measurable converging to f. If $|f_k| \le \varphi$ for any $k \in \mathbb{N}$ and $\varphi \in L(X, \mathscr{A}, \mu)$, then*

$$\int_X f_k d\mu \longrightarrow \int_X f d\mu \quad (k \to \infty).$$

Theorem 1.2.10. *Suppose $f \in (X, \mathscr{A}, \mu)$, then for any $\varepsilon > 0$, there exists a $\delta > 0$ such that for all $A \in \mathscr{A}$ with $\mu(A) < \delta$, we have*

$$\int_X |f| \chi_A d\mu < \varepsilon.$$

Remark. From now on, $\int_X f \chi_A d\mu$ will be written as $\int_A f d\mu$.

Exercise 1.2

1. Suppose $(X, \mathscr{A}, \mu)$ is a complete measure space. Prove that if f is measurable and $f = g$ μ-a.e., then g is measurable. If $L(X, \mathscr{A}, \mu)$ is not complete, is this proposition still true? Please give an example.
2. Show Theorems 1.2.1–1.2.10, especially Theorems 1.2.7 and 1.2.10.

1.3 $L^p(X, \mathscr{A}, \mu)$

Definition 1.3.1. Let $(X, \mathscr{A}, \mu)$ be a measure space, f be a $\mathscr{A}$-measurable generalized real-valued function on X and $0 < p < \infty$. Define

$$L^p(X, \mathscr{A}, \mu) = \{f : |f|^p \in L(X, \mathscr{A}, \mu)\},$$

$$\|f\|_p = \|f\|_{L^p(X, \mathscr{A}, \mu)} = \left\{ \int_X |f|^p d\mu \right\}^{\frac{1}{p}},$$

$$\|f\|_\infty = \inf_{\mu(E)=0} \sup\{|f(x)| : x \in X \backslash E\},$$

$$L^\infty(X, \mathscr{A}, \mu) = \{f : \|f\|_\infty < \infty\}.$$

Remark. According to Theorem 1.2.5(3), those functions in $L^p(X, \mathscr{A}, \mu)$ are considered as the same element if they are equal a.e. Thus, it can be assumed that the elements of $L^p(X, \mathscr{A}, \mu)$ are functions that take finite values only.

Lemma 1.3.1 (Young's inequality). *Let $\varphi \in \mathbb{C}_{[0,\infty)}$ be a strictly increasing function with $\varphi(0) = 0$. If $\psi(t) = \varphi^{-1}(t)$,*

$$ab \leq \int_0^a \varphi(t)dt + \int_0^b \psi(t)dt \quad \text{for all } a, b > 0.$$

Proof. Without loss of generality, assume $a \geq \varphi^{-1}(b) = \psi(b)$. The variable substitution of Riemann–Stieltjes integral indicates

$$\int_0^b \psi(t)dt \xrightarrow{t=\varphi(u)} \int_0^{\varphi^{-1}(b)} \psi(\varphi(u))d\varphi(u)$$

$$= \int_0^{\varphi^{-1}(b)} u \, d\varphi(u) = \varphi^{-1}(b)b - \int_0^{\varphi^{-1}(b)} \varphi(u)du.$$

Moreover, we obtain

$$\int_0^a \varphi(t)dt + \int_0^b \psi(t)dt = \varphi^{-1}(b)b + \int_{\varphi^{-1}(b)}^a \varphi(u)du$$

$$\geq \varphi^{-1}(b)b + b(a - \varphi^{-1}(b)) = ab,$$

where the equality holds if and only if $a = \varphi^{-1}(b)$. $\square$

Theorem 1.3.2 (Hölder's inequality). *Let*

$$1 \leq p \leq \infty, \frac{1}{p} + \frac{1}{p'} = 1$$

and regard $\frac{1}{\infty}$ *as 0. If* $f \in L^p(X, \mathscr{A}, \mu)$ *and* $g \in L^{p'}(X, \mathscr{A}, \mu)$, *then* $fg \in L(X, \mathscr{A}, \mu)$ *and*

$$\int_X |fg|d\mu \leq \|f\|_p \|g\|_{p'}.$$

Proof. Consider the case of $1 < p < \infty$ first. Take

$$\varphi(t) = t^{p-1} \quad (0 \leq t < \infty),$$

then its inverse function is $\psi(t) = t^{p'-1}$. Using Young's inequality,

$$|f(x)g(x)| \leq \int_0^{|f(x)|} t^{p-1}dt + \int_0^{|g(x)|} t^{p'-1}dt$$

$$= \frac{1}{p}|f(x)|^p + \frac{1}{p'}|g(x)|^{p'},$$

$$\int_X |fg|d\mu \leq \frac{1}{p}\|f\|_p^p + \frac{1}{p'}\|g\|_{p'}^{p'}.$$

Replace f with $f/\|f\|_p$ and g with $g/\|g\|_{p'}$ in the above equation, and note that $\frac{1}{p} + \frac{1}{p'} = 1$ leads to the desired conclusion.

For $p = 1$ and $p' = \infty$, it is clear that

$$\int_X |fg| d\mu \le \|g\|_\infty \|f\|_1.$$

$\square$

Theorem 1.3.3 (Minkowski's inequality). *Let* $1 \le p \le \infty$. *If* $f, g \in L^p(X, \mathscr{A}, \mu)$, *then*

$$\|f + g\|_p \le \|f\|_p + \|g\|_p.$$

Proof. If $p = 1$ or $p = \infty$, then it is obvious that

$$\|f + g\|_p \le \|f\|_p + \|g\|_p$$

from $|f + g| \le |f| + |g|$. Besides, consider $1 < p < \infty, p' = \frac{p}{p-1}$,

$$|f + g|^p = |f + g||f + g|^{p-1} \le |f + g|^{p-1}|f| + |f + g|^{p-1}|g|.$$

Hölder's inequality gives

$$\int_X |f||f + g|^{p-1} d\mu \le \|f\|_p \cdot \|f + g\|_p^{p/p'},$$

$$\int_X |g||f + g|^{p-1} d\mu \le \|g\|_p \cdot \|f + g\|_p^{p/p'}.$$

Then

$$\int_X |f + g|^p d\mu \le (\|f\|_p + \|g\|_p) \|f + g\|_p^{p/p'},$$

that is,

$$\|f + g\|_p \le \|f\|_p + \|g\|_p.$$

$\square$

Theorem 1.3.4. *Let* $1 \le p \le \infty$. *A space* $L^p(X, \mathscr{A}, \mu)$ *is a real Banach space in norm* $\| \cdot \|_p$.

Proof.   Theorem 1.3.3 tells us that $(L^p, \|\cdot\|_p)$ is a real normed linear space. Now, prove its completeness. Suppose $\{f_n\}_{n=1}^{\infty}$ is a Cauchy sequence in $(L^p, \|\cdot\|_p)$ and consider the case $p = \infty$ first. Let

$$E_{n,k} = \{x \in X : |f_n(x) - f_k(x)| > \|f_n - f_k\|_\infty\}, \quad n, k \in \mathbb{N}$$

and $E = \cup_{n,k\in\mathbb{N}}E_{n,k}$. The definition of $\|\cdot\|_\infty$ indicates $\mu(E_{n,k}) = 0$, thus $\mu(E) = 0$ and

$$|f_n(x) - f_k(x)| \leq \|f_n - f_k\|_\infty \to 0 \quad (n, k \to \infty)$$

on $X \backslash E$. It follows that there exists

$$\lim_{n\to\infty} f_n(x) =: f(x)$$

uniformly on $\complement E$. It may be assumed that f takes zero on E. Then $f \in L^\infty$ and

$$\|f - f_n\|_\infty \to 0 \quad (n \to \infty).$$

Next, consider $1 \leq p < \infty$. Take a subsequence $\{g_k\}_{k=1}^{\infty}$ such that

$$\|g_k - g_{k+1}\|_p < 2^{-k}, \quad k \in \mathbb{N}.$$

One might as well assume that g_k take finite values only. Let

$$g = |g_1| + \sum_{k=1}^{\infty} |g_{k+1} - g_k|.$$

By the Levi theorem and Minkowski's inequality,

$$\int_X g^p d\mu = \int_X \left(|g_1| + \sum_{k=1}^{\infty} |g_{k+1} - g_k|\right)^p d\mu$$

$$\leq \lim_{n\to\infty} \left(\|g_1\|_p + \sum_{k=1}^{n} \|g_{k+1} - g_k\|_p\right)^p$$

$$\leq (\|g_1\|_p + 1)^p < \infty.$$

It follows that g is finite μ-a.e. Therefore,

$$\lim_{n\to\infty} g_n = g_1 + \sum_{k=1}^{\infty} (g_{k+1} - g_k) =: f$$

is finite μ-a.e. (the definition of f can be supplemented at points where g_n does not converge such that f takes zero at these points).

Thus,

$$|f| \le g \in L^p.$$

For any given $\varepsilon > 0$, take N to be large enough such that

$$\|g_m - g_n\|_p < \varepsilon$$

when $m, n \ge N$. Applying Fatou's lemma, let $n \to \infty$ to get

$$\|g_m - f\|_p \le \varepsilon, \quad m > N.$$

It can be seen that $\lim_{k \to \infty} \|g_k - f\|_p = 0$. Thus, $\lim_{n \to \infty} \|f_n - f\|_p = 0$. $\square$

In order to estimate the L^p norm, it is necessary to introduce distribution function. Let $(X, \mathscr{A}, \mu)$ be a measure space and f be a finite μ-a.e. and measurable function. The distribution function of f is defined by

$$\lambda(t) = \mu(\{x \in X : f(x) > t\}), \quad t \in \mathbb{R}.$$

Clearly, $\lambda(t)$ is decreasing and right continuous. By extension, λ is said to be right continuous as long as $\lim_{t \to t_0^+} \lambda(t) = \lambda(t_0)$, regardless of whether $\lambda(t_0)$ is finite or not.

Theorem 1.3.5. *Let $(X, \mathscr{A}, \mu)$ be a measure space, f be a nonnegative, finite μ-a.e. and measurable function, and λ be the distribution function of f. Then*

$$\int_X f d\mu = \int_0^\infty \lambda(t) dt,$$

where $\int_0^\infty \lambda(t) dt$ is a usual Lebesgue integral.

Proof.   Divide the proof into five steps:

(1) It can be argued that f takes positive values everywhere.
(2) Let $E_{n,k} = \{x \in X : \frac{k-1}{2^n} < f(x) \le \frac{k}{2^n}\}$ and define

$$f_n(x) = \frac{k-1}{2^n}, \quad \text{if } x \in E_{n,k}, k \in \mathbb{N}.$$

It is obvious that

$$X = \bigcup_{k=1}^{\infty} E_{n,k}$$

is a disjoint union. If $x \in E_{n,k}$, then x either belongs to $E_{n+1,2k-1}$ or $E_{n+1,2k}$ and

$$0 \le f_n(x) \le f_{n+1}(x) \to f(x)$$

since

$$E_{n,k} = E_{n+1,2k-1} \cup E_{n+1,2k}.$$

Then, by the Levi theorem,

$$\int_X f d\mu = \lim_{n \to \infty} \int_X f_n d\mu.$$

(3) Denote the integral of f_n by λ. Note that

$$\sum_{k=1}^{N} \frac{k-1}{2^n} \mu(E_{n,k}) = \sum_{k=2}^{N} \frac{1}{2^n} \cdot \sum_{j=k}^{N} \mu(E_{n,j})$$

$$= \sum_{k=2}^{N} \frac{1}{2^n} \mu \left(\bigcup_{j=k}^{N} E_{n,j} \right)$$

$$= \sum_{k=2}^{N} \frac{1}{2^n} \mu \left(\left\{ x \in X : \frac{k-1}{2^n} < f(x) \le \frac{N}{2^n} \right\} \right).$$

Then

$$\sum_{k=1}^{\infty} \frac{k-1}{2^n} \mu(E_{n,k}) = \sum_{k=2}^{\infty} \frac{1}{2^n} \lambda \left(\frac{k-1}{2^n} \right).$$

Therefore,

$$\int_X f_n d\mu = \sum_{k=1}^{\infty} \int_{E_{n,k}} f_n d\mu = \sum_{k=1}^{\infty} \frac{k-1}{2^n} \mu(E_{n,k})$$

$$= \sum_{k=2}^{\infty} \frac{1}{2^n} \lambda \left(\frac{k-1}{2^n} \right).$$

(4) Suppose $\varphi_n(t) = \lambda\left(\frac{k}{2^n}\right)$. If $t \in \left(\frac{k-1}{2^n}, \frac{k}{2^n}\right], k \in \mathbb{N}$, then

$$0 \leq \varphi_n(t) \leq \varphi_{n+1}(t) \leq \lambda(t).$$

For any $t > 0$, there exists a unique $t_n = \frac{k}{2^n}$ such that $t \leq t_n < t + \frac{1}{2^n}$. When $n \to \infty$, $t_n \to t^+$. Since λ is right continuous,

$$\varphi_n(t) = \lambda(t_n) \to \lambda(t).$$

Then we have

$$\int_X f_n d\mu = \sum_{k=2}^{\infty} \frac{1}{2^n} \lambda\left(\frac{k-1}{2^n}\right) = \sum_{k=1}^{\infty} \int_{(k-1)/2^n}^{k/2^n} \varphi_n(t) dt$$
$$= \int_0^{\infty} \varphi_n(t) dt,$$

which is a usual Lebesgue integral.

(5)
$$\int_X f d\mu = \lim_{n \to \infty} \int_X f_n d\mu = \lim_{n \to \infty} \int_0^{\infty} \varphi_n(t) dt = \int_0^{\infty} \lambda(t) dt.$$

$\square$

Corollary 1.3.6. *Suppose* $f \in L^p(X, \mathscr{A}, \mu), 0 < p < \infty$ *and* λ *is the distribution function of* $|f|$, *then*

$$\int_X |f|^p d\mu = p \int_0^{\infty} t^{p-1} \lambda(t) dt.$$

Proof. For any $t \geq 0$,

$$\{x \in X : |f|^p > t\} = \{x \in X : |f| > t^{\frac{1}{p}}\}.$$

According to Theorem 1.3.5, we have

$$\int_X |f|^p d\mu = \int_0^{\infty} \lambda(t^{\frac{1}{p}}) dt.$$

Then after the change of variable $t = s^p$, the desired result is obtained. $\square$

Example 1.3.1. Let D be a nonempty collection and f be a real-valued function on D. For $0 < p < \infty$, define

$$l_p(f) = \left\{ \sum_{x \in D} |f(x)|^p \right\}^{\frac{1}{p}}$$

$$:= \sup \left\{ \left[\sum_{x \in F} |f(x)|^p \right]^{\frac{1}{p}} : F \text{ is a finite subset of } D \right\}.$$

Suppose $\mathscr{A} = 2^D$ and μ is counting measure on $\mathscr{A}$, then

$$L^p(D, \mathscr{A}, \mu) = \{f : l_p(f) < \infty\} =: l^p(D)$$

and

$$\|f\|_p = l_p(f).$$

We need to point out that if $l_p(f) < \infty$, then there exists a countable set $M \subset D$ such that $f(x) = 0$ when $x \in D \backslash M$, at this time,

$$l_p(f) = \left(\sum_{x \in M} |f(x)|^p \right)^{\frac{1}{p}}.$$

Therefore, $l^p(D)$ constitutes a Banach space in norm l_p when $1 \leq p < \infty$. If $D = \mathbb{N}$, we abbreviate $l^p(\mathbb{N})$ as l^p, which is well known (refer to any of the Functional Analysis textbooks).

The case of $p = 2$ plays a crucial role in the L^p theory. Therefore, we have some discussion about $L^2(X, \mathscr{A}, \mu)$ below. Introduce an inner product on real $L^2(X, \mathscr{A}, \mu)$:

$$\langle f, g \rangle = \int_X fg d\mu,$$

then $L^2(X, \mathscr{A}, \mu)$ becomes a Hilbert space based on it. Here, we discuss a little bit about the Hilbert space.

Theorem 1.3.7. *Let H be a Hilbert space and E be an orthonormal basis for H. The following are equivalent:*

(1) *E is complete, that is, $x \perp E \Rightarrow x = 0$.*
(2) *For each $x \in H$, there exists a countable set $I \subset E$ such that $x \perp (E \backslash I)$ and*

$$x = \sum_{z \in I} \langle x, z \rangle z.$$

(3) *For each $x \in H$, $\|x\|^2 = \sum_{z \in E} |\langle x, z \rangle|^2$ (Parseval's identity).*
(4) *For all $x, y \in H$, $\langle x, y \rangle = \sum_{z \in E} \langle x, z \rangle \langle z, y \rangle$ (Parseval's identity).*
(5) *The smallest subspace of H containing E is dense in H.*

Proof. Suppose (1) holds and $x \in H$. For any finite number of elements $z_1, \ldots, z_k \in E$, denote

$$u = \sum_{j=1}^{k} \langle x, z_j \rangle z_j,$$

then

$$0 \leq \langle x - u, x - u \rangle = \|x\|^2 + \|u\|^2 - \langle x, u \rangle - \langle u, x \rangle.$$

Note that

$$\|u\|^2 = \sum_{j=1}^{k} |\langle x, z_j \rangle|^2 = \langle x, u \rangle,$$

we have

$$\sum_{j=1}^{k} |\langle x, z_j \rangle|^2 \leq \|x\|^2.$$

It follows that there exist at most countable elements in E that are not orthogonal to x (refer to Exercise 8), which is denoted by $\{z_n\}$. Then

$$y := \sum_{n} \langle x, z_n \rangle z_n \in H$$

and $x - y \perp E$. Thus, $x = y$ and (2) is obtained.

Suppose (2) is true. For any $x, y \in H$, there exists a sequence $\{z_n\}$ such that

$$x = \sum_{n} \langle x, z_n \rangle z_n, \quad y = \sum_{n} \langle y, z_n \rangle z_n.$$

Denote $x_n = \sum_{k=1}^{n} \langle x, z_k \rangle z_k$ and $y_n = \sum_{k=1}^{n} \langle y, z_k \rangle z_k$, then $x_n \xrightarrow{\|\cdot\|} x$ and $y_n \xrightarrow{\|\cdot\|} y$. Thus,

$$
\begin{aligned}
|\langle x, y \rangle - \langle x_n, y_n \rangle| &= |\langle x - x_n, y \rangle + \langle x_n, y - y_n \rangle| \\
&\leq \|x - x_n\| \, \|y\| + \|x_n\| \, \|y - y_n\| \\
&\leq \|x - x_n\|(\|y\| + \|y - y_n\|) + \|y - y_n\| \\
&\|x\| \to 0, \ n \to \infty.
\end{aligned}
$$

Then (4) can be obtained by applying

$$
\langle x_n, y_n \rangle = \sum_{k=1}^{n} \langle x, z_k \rangle \overline{\langle y, z_k \rangle}.
$$

Obviously, (4) $\Rightarrow$ (3) $\Rightarrow$ (1), that is, (1), (2), (3) and (4) are equivalent to each other.

Finally, we focus on (5). It is easy to show (2) $\Rightarrow$ (5). Now, suppose (5) holds and $x \perp E$, then $x \perp \mathrm{span}\, E$ and there exists a $y_n \in \mathrm{span}\, E$ such that $\|y_n - x\| \to 0$ $(n \to \infty)$. Consequently,

$$
\langle x, x - y_n \rangle = \|x\|^2 \leq \|x\| \, \|x - y_n\| \to 0.
$$

Thus, $x = 0$ and (1) is proved. $\qquad\square$

Theorem 1.3.8. *Any two complete orthonormal bases of a Hilbert space have the same cardinality.*

Proof. Let A and B be complete orthonormal bases for a Hilbert space H. If A is a finite set, then A is a Hamel basis for H. Inasmuch as B is linearly independent, $\mathrm{card}(B) \leq \mathrm{card}(A)$ and B is also a Hamel basis. Thus, $\mathrm{card}(A) = \mathrm{card}(B)$. Next, suppose A and B are infinite sets. For all $a \in A$, define $B_a = \{b \in B : \langle a, b \rangle \neq 0\}$, then $\mathrm{card}(B_a) \leq \aleph_0$. Given $b \in B$, there must be some $a \in A$ such that $\langle b, a \rangle \neq 0$, i.e. $b \in B_a$, for

$$
1 = \|b\|^2 = \sum_{a \in A} |\langle b, a \rangle|^2.
$$

Hence,

$$
B = \bigcup_{a \in A} B_a.
$$

From this, it can be concluded that $\mathrm{card}(B) \leq \aleph_0 \, \mathrm{card}(A) = \mathrm{card}(A)$. Swapping the positions of A and B yields $\mathrm{card}(A) = \mathrm{card}(B)$. $\qquad\square$

Theorem 1.3.9. *If H is a Hilbert space, then H must possess orthonormal bases.*

Proof.   It is obvious that every identity element of H forms a single-element orthonormal basis. All orthonormal bases for H are taken as a family $\mathscr{F}$. Define a partial ordering on $\mathscr{F}$: for $A, B \in \mathscr{F}$, $A \leq B$ provided $A \subset B$. Every totally ordered subset of this partial ordering has upper bounds (if $\mathscr{D} = \{A\}$ is a totally ordered subset of $\mathscr{F}$, then $M := \cup_{A \in \mathscr{D}} A$ is an upper bound for $\mathscr{D}$). Using the Zorn lemma, it can be inferred that $\mathscr{F}$ has maximal members, which is denoted by M. M must be complete. Otherwise, it is known from Theorem 1.3.7(1) that there exists an $x \in H$, $x \neq 0$ and $x \perp M$. Merge $\frac{x}{\|x\|}$ into M to become a member of $\mathscr{F}$, which is greater than M and contradicts M as a maximal element.   □

Definition 1.3.2. Let H be a Hilbert space. The cardinality of the complete orthonormal basis of H is called the orthogonal dimension of H or dimension for short.

Theorem 1.3.10. *Let H be a Hilbert space, then there exists a set D such that H and $l^2(D)$ are isometric isomorphic.*

Proof.   Take D as a complete orthonormal basis for H. For any $x \in H$, define f_x:

$$f_x(z) = \langle x, z \rangle \quad \text{for all } z \in D.$$

Obviously, $f_x \in l^2(D)$ (see Example 1.3.1) and

$$l_2(f_x) = \left\{ \sum_{z \in D} |f_x(z)|^2 \right\}^{\frac{1}{2}} = \left\{ \sum_{z \in D} |\langle x, z \rangle|^2 \right\}^{\frac{1}{2}}.$$

Note that the mapping $T : x \mapsto f_x$ is linear and in Theorem 1.3.7(4), for any $x, y \in H$,

$$\langle x, y \rangle = \sum_{z \in D} \langle x, z \rangle \overline{\langle y, z \rangle} = \sum_{z \in D} f_x(z) \cdot \overline{f_y(z)} := \langle f_x, f_y \rangle.$$

It is clear that the mapping T preserves the inner product.

Let $f \in l^2(D)$ and $l_2(f) = \left(\sum_{z \in D} |f(z)|^2 \right)^{\frac{1}{2}}$. Define

$$x = \sum_{z \in D} f(z) z.$$

Observe that this series converges in the H norm, so the definition is reasonable. Hence, $x \in H$ and $\langle x, z \rangle = f(z)$. It can be seen that $T(x) = f$, thus T is surjective. Combining that T is injective, we prove that under the mapping T, H and $l^2(D)$ are isometric isomorphic. $\qquad \square$

Theorem 1.3.11. *Let D be a nonempty set. Then the bounded linear functional L on $l^2(D)$ can be represented as an inner product: There exists a unique $g \in l^2(D)$ such that $\|L\| = l_2(g)$ and*

$$L(f) = \langle f, g \rangle, \quad \text{for all } f \in l^2(D).$$

Proof. Let $\xi \in D$. Define

$$\chi_\xi(x) = \begin{cases} 1 & \text{if } x = \xi, \\ 0 & \text{if } x \neq \xi, x \in D. \end{cases}$$

Then $\chi_\xi \in l^2(D)$. Note $L(\chi_\xi) = a_\xi$. Take $\alpha_1, \ldots, \alpha_k \in \mathbb{C}$ and pairwise different $\xi_1, \ldots, \xi_k \in D$ and suppose

$$f = \sum_{i=1}^{k} \alpha_i \chi_\xi,$$

whereupon

$$|L(f)| = \left| \sum_{j=1}^{k} \alpha_j \cdot a_{\xi_j} \right| \leq \|L\| \cdot l_2(f) = \|L\| \cdot \left(\sum_{j=1}^{k} |\alpha_j|^2 \right)^{\frac{1}{2}}.$$

If $\alpha_j = \bar{a}_{\xi_j}$, then

$$\left(\sum_{j=1}^{k} |a_{\xi_j}|^2 \right)^{\frac{1}{2}} \leq \|L\|,$$

hence

$$\left\{ \sum_{\xi \in D} |L(\chi_\xi)|^2 \right\}^{\frac{1}{2}} \leq \|L\|.$$

Define

$$g(\xi) = \overline{L(\chi_\xi)}, \quad \xi \in D,$$

then $g \in l^2(D)$ and $l^2(g) \leq \|L\|$. For $f \in l^2(D)$,

$$f = \sum_{\xi \in D} f(\xi)\chi_\xi$$

in the $l^2(D)$ norm. As a result,

$$L(f) = \sum_{\xi \in D} f(\xi)L(\chi_\xi) = \sum_{\xi \in D} f(\xi)\, \overline{g(\xi)},$$

the right end is exactly $\langle f, g \rangle$ by definition, thereby

$$L(f) = \langle f, g \rangle.$$

By Hölder's inequality, we obtain

$$|L(f)| = l_2(f) \cdot l_2(g).$$

Thus, $\|L\| \leq l_2(g)$, and therefore, $\|L\| = l_2(g)$.   □

The following corollary is obtained from Theorems 1.3.10 and 1.3.11.

Corollary 1.3.12. *Let H be a Hilbert space and L be a bounded linear functional. Then there exists a unique $g \in H$ such that*

$$L(f) = \langle f, g \rangle \quad \text{for all } f \in H.$$

Theorem 1.3.11 and Corollary 1.3.12 are called the representation theorems named after Riesz.

Exercise 1.3

1. Let $0 < p < 1$. Define a binary function d on $L^p(X, \mathscr{A}, \mu)$:

$$d(f, g) = \int_X |f - g|^p d\mu, \quad f, g \in L^p(X, \mathscr{A}, \mu).$$

Show that (L^p, d) is a complete metric space.

2. Let $f \in L^\infty(X, \mathscr{A}, \mu)$. Prove that

$$\|f\|_\infty = \inf\{\alpha > 0 : \mu(\{x \in X : |f(x)| > \alpha\}) = 0\}.$$

3. If $f \in L^p(X, \mathscr{A}, \mu)$ for any $p \in [1, \infty)$, then

$$\|f\|_\infty = \lim_{p \to \infty} \|f\|_p.$$

4. Suppose μ is a usual Lebesgue measure on $\mathbb{R}^n$, $p, q \in (0, \infty)$ and $p \neq q$. Please seek an $f \in L^p(\mathbb{R}^n) \backslash L^q(\mathbb{R}^n)$.

5. The Vitali convergence theorem: Suppose $p \in [1, \infty)$ and $\{f_n\} \subset L^p(X, \mathscr{A}, \mu)$ with $f_n \to f$ and f is finite μ-a.e. In order for $f \in L^p(X, \mathscr{A}, \mu)$ and $\lim_{n \to \infty} \|f - f_n\|_p = 0$, it is sufficient for the following two conditions to hold:

   (1) For any $\varepsilon > 0$, there exists an $A_\varepsilon \in \mathscr{A}, \mu(A_\varepsilon) < \infty$ such that

   $$\int_{X \backslash A_\varepsilon} |f_n|^p d\mu < \varepsilon \quad \text{for all } n \in \mathbb{N};$$

   (2) For n,

   $$\lim_{\mu(E) \to 0} \int_E |f_n|^p d\mu = 0$$

   holds uniformly.

6. The orthogonal dimension of $L^2(\mathbb{R}^n)$, a Hilbert space, is $\aleph_0$.

7. Let d be the orthogonal dimension of the Hilbert space H. Prove H is separable if and only if $d \leq \aleph_0$.

8. Let E be a set of certain positive numbers. If $\text{card}(E) > \aleph_0$, then a countable subset $\{a_n : n \in \mathbb{N}\}$ can be extracted from E, such that

$$\sum_{n=1}^{\infty} a_n = \infty.$$

1.4 Signed Measure

In this section, we discuss the decomposition for signed measures and its relationship with measures, introducing the concepts of absolute continuity and singularity.

1.4.1   The Decomposition for Signed Measure

Definition 1.4.1. Let φ be a signed measure on a measurable space $(X, \mathscr{A})$. A set $P \in \mathscr{A}$ is called a positive set with respect to φ provided

$$\varphi(P \cap E) \geq 0 \quad \text{for all } E \in \mathscr{A}.$$

A set is called a negative set with respect to φ provided it is a positive set with respect to $-\varphi$. A measurable set is called a null set with respect to φ provided it is positive and negative.

Lemma 1.4.1. *Let φ be a signed measure on a measurable space $(X, \mathscr{A})$, then*

(1) *if $E, F \in \mathscr{A}$ satisfy $\varphi(E) > -\infty$ and $F \subset E$, then*

$$\varphi(F) > -\infty;$$

(2) *if $A_n \in \mathscr{A}$ and $A_n \nearrow A$, then*

$$\lim_{n \to \infty} \varphi(A_n) = \varphi(A);$$

(3) *if $B_n \in \mathscr{A}$ and $B_n \searrow B$ with $|\varphi(B_1)| < \infty$, then*

$$\lim_{n \to \infty} \varphi(B_n) = \varphi(B).$$

Proof. The fact $\varphi(E) = \varphi(F) + \varphi(E \backslash F)$ implies (1). The proofs of (2) and (3) are the same as the proofs of Theorems 1.1.2 and 1.1.3, respectively. $\qquad\square$

Theorem 1.4.2. *Let φ be a signed measure on a measurable space $(X, \mathscr{A})$ and $E \in \mathscr{A}$. If $\varphi(E) > 0$, then there exists a set $S \in \mathscr{A}$ such that $S \subset E$ with $\varphi(S) > 0$ and S is positive with respect to φ.*

Proof. According to Lemma 1.4.1 (1), for every measurable set $F \subset E$, $\varphi(F) > -\infty$. If E is not positive, then there exists a natural number

$$n_1 = \min\left\{ k \in \mathbb{N} : \text{ there exists a set } A \in \mathscr{A}, A \subset E, \varphi(A) < -\frac{1}{k} \right\},$$

$F_1 \in \mathscr{A}$ and $F_1 \subset E$ such that $\varphi(F_1) < -\frac{1}{n_1}$. Thus,

$$\varphi(F_1) \in \left(-\infty, -\frac{1}{n_1}\right)$$

and

$$\varphi(E \backslash F_1) = \varphi(E) - \varphi(F_1) \geq \varphi(E) + \frac{1}{n_1} > 0.$$

If $E \backslash F_1$ is a positive set, then Theorem 1.4.2 has been established. Otherwise, there exists another natural number

$$n_2 = \min \Big\{ k \in \mathbb{N} : \text{there exists a set}$$

$$A \in \mathscr{A}, A \subset E \backslash F_1, \varphi(A) < -\frac{1}{k} \Big\},$$

$F_2 \in \mathscr{A}$ and $F_2 \subset E \backslash F_1$ such that $-\infty < \varphi(F_2) < -\frac{1}{n_2}$. Thus, $n_2 \geq n_1$ and

$$\varphi((E \backslash F_1) \backslash F_2) > \varphi(E) + \frac{1}{n_1} + \frac{1}{n_2} > 0.$$

If $(E \backslash F_1) \backslash F_2$ is a positive set, then Theorem 1.4.2 has been established. Otherwise, repeat the above steps for $(E \backslash F_1) \backslash F_2$.

Continuing to do so will result in two outcomes. One result is to obtain a positive set

$$E \Big\backslash \bigcup_{j=1}^{m} F_j = S,$$

such that $\varphi(S) > 0$ after a finite number of iterations, thereby the conclusion of Theorem 1.4.2 is confirmed. The other result is that these steps are endless, resulting in a sequence of pairwise disjoint measurable sets $F_k, k \in \mathbb{N}$ satisfying $F_k \subset E$ and a monotonic increasing subsequence of natural numbers $\{n_k\}_{k=1}^{\infty}$ such that

$$n_k = \min \Big\{ j \in \mathbb{N} : \text{there exists a set}$$

$$\times \ A \in \mathscr{A}, A \subset E \setminus \bigcup_{i=1}^{k-1} F_i, \varphi(A) < -\frac{1}{j} \Big\},$$

$$-\infty < \varphi(F_k) < -\frac{1}{n_k},$$

where it is specified that $\cup_{i=1}^{\infty} F_i = \emptyset$. Let $S = E \setminus \cup_{k=1}^{\infty} F_k$, then

$$\varphi(S) \geq \varphi(E) + \sum_{k=1}^{\infty} \frac{1}{n_k} > 0.$$

Due to the fact

$$-\infty < \varphi\left(\bigcup_{k=1}^{\infty} F_k\right) = \sum_{k=1}^{\infty} \varphi(F_k) \leq -\sum_{k=1}^{\infty} \frac{1}{n_k},$$

$\lim_{k\to\infty} n_k = \infty$. Bearing the definition of n_k in mind, we obtain that for any $A \in \mathscr{A}$ with $A \subset S$,

$$\varphi(A) \geq -\frac{1}{n_k - 1} \quad \text{if } n_k \geq 2.$$

It is thus clear that $\varphi(A) \geq 0$, that is, S is positive. $\qquad\square$

Theorem 1.4.3 (Hahn's decomposition theorem). *Let φ be a signed measure on a measurable space $(X, \mathscr{A})$, then there exists a set $P \in \mathscr{A}$ such that P is positive with respect to φ and $\complement P = X \backslash P$ is negative with respect to φ. The decomposition is called Hahn decomposition, abbreviated as $(P, \complement P)$, which is unique in the following meaning: If there exists another Hahn decomposition $(P_1, \complement P_1)$, then for any $A \in \mathscr{A}$, we have*

$$\varphi(P_1 \cap A) = \varphi(P \cap A), \quad \varphi(\complement P_1 \cap A) = \varphi(\complement P \cap A).$$

Proof. Without loss of generality, we assume that φ does not take the value ∞, that is, for any $A \in \mathscr{A}$, $\varphi(A) < \infty$. Let

$$\alpha = \sup\{\varphi(A) : \ A \text{ is positive}\}.$$

Take a sequence $\{A_k\}_{k=1}^{\infty}$ of positive sets such that

$$\lim_{k \to \infty} \varphi(A_k) = \alpha.$$

Assuming

$$P_n = \bigcup_{k=1}^{n} A_k,$$

it is obvious that P_n is positive and

$$\alpha \geq \varphi(P_{n+1}) \geq \varphi(P_n) \geq \varphi(A_n).$$

Let

$$P = \bigcup_{k=1}^{\infty} A_k.$$

By Lemma 1.4.1 (2),

$$\varphi(P) = \lim_{n \to \infty} \varphi(P_n) = \alpha < \infty.$$

Suppose that $\complement P$ is not negative, then there exists an $E \in \mathscr{A}$ with $E \subset \complement P$ such that $\varphi(E) > 0$. According to Theorem 1.4.2, it is known that there exists a positive set $S \in \mathscr{A}$ such that $S \subset E$ with $\varphi(S) > 0$. Then $S \cup P$ is positive and

$$\varphi(S \cup P) = \varphi(S) + \varphi(P) > \alpha$$

which contradicts the definition of α. Hence, $\complement P$ is negative.

Assuming there exists another positive set P_1 and its corresponding negative set $\complement P_1$, we can obtain that $P \triangle P_1 = (P \cap \complement P_1) \cup (P_1 \cap \complement P)$ must be null, thus $\complement P \triangle \complement P_1$ is also null. Thereby, the Hahn decomposition is unique. $\qquad \square$

Definition 1.4.2. Let φ be a signed measure on a measurable space $(X, \mathscr{A})$. For any $E \in \mathscr{A}$,

$$\varphi^+(E) = \sup_{A \in \mathscr{A}} \varphi(A \cap E),$$

$$\varphi^-(E) = \sup_{A \in \mathscr{A}} \{-\varphi(A \cap E)\},$$

$$|\varphi|(E) = \varphi^+(E) + \varphi^-(E)$$

are called positive, negative and total variations of φ on E, respectively.

Theorem 1.4.4. *Let φ be a signed measure on a measurable space $(X, \mathscr{A})$. If P, $\complement P$ are positive and negative with respect to φ, respectively, then for any $E \in \mathscr{A}$,*

$$\varphi^+(E) = \varphi(E \cap P),$$
$$\varphi^-(E) = -\varphi(E \cap \complement P).$$

Proof. For any measurable subset F of E,

$$\varphi(F) = \varphi(F \cup P) + \varphi(F \cap \complement P) \le \varphi(F \cap P) \le \varphi(E \cap P),$$
$$\varphi(F) \ge \varphi(F \cap \complement P) \ge \varphi(E \cap \complement P).$$

$\square$

Theorem 1.4.5 (Jordan's decomposition theorem). *Let φ be a signed measure on a measurable space $(X, \mathscr{A})$, then*

$$\varphi = \varphi^+ - \varphi^-.$$

Proof. According to Theorem 1.4.3, we can assume $(P, \complement P)$ is a Hahn decomposition. By Theorem 1.4.4, for any $E \in \mathscr{A}$,

$$\varphi(E) = \varphi(E \cap P) + \varphi(E \cap \complement P) = \varphi^+(E) - \varphi^-(E).$$

$\square$

The Jordan decomposition theorem tells us that the signed measure must be the difference between two measures, its positive variation and negative variation. If $f \in L(X, \mathscr{A}, |\varphi|)$, then define

$$\int_X f d\varphi = \int_X f d\varphi^+ - \int_X f d\varphi^-.$$

1.4.2 Absolute Continuity and Singularity

Definition 1.4.3. Let $(X, \mathscr{A})$ be a measurable space, μ and ν be signed measures. We say ν is absolutely continuous with respect to μ, denoted by $\nu \ll \mu$ if $\nu(E) = 0$, for any $E \in \mathscr{A}$, for which $|\mu|(E) = 0$.

Example 1.4.1. Suppose $(X, \mathscr{A}, \mu)$ is a measure space and $f \in L(X, \mathscr{A}, \mu)$. Let

$$\nu(A) = \int_A f d\mu \quad \text{for any } A \in \mathscr{A},$$

then $\nu \ll \mu$.

Definition 1.4.4. Let $(X, \mathscr{A})$ be a measurable space. We say that two signed measures μ and ν are mutually singular or that μ is singular with respect to ν, which is expressed by $\mu \perp \nu$, if there exists an $A \in \mathscr{A}$ such that

$$|\mu|(A) = |\nu|(\complement A) = 0.$$

Example 1.4.2. Suppose $(X, \mathscr{A}, \mu)$ is a measure space, ψ is a signed measure on $(X, \mathscr{A})$ and $E \in \mathscr{A}$ with $\mu(E) = 0$. Let

$$\varphi(A) = \psi(A \cap E) \quad \text{for any } A \in \mathscr{A},$$

then $\varphi \perp \mu$.

Theorem 1.4.6. *Let $(X, \mathscr{A}, \mu)$ be a measure space and ν be a finite signed measure on $(X, \mathscr{A})$. Then $\nu \ll \mu$ if and only if for any $\varepsilon > 0$, there exists a $\delta > 0$ such that, for every $A \in \mathscr{A}$, $|\nu|(A) < \varepsilon$ provided $\mu(A) < \delta$.*

Proof. On the one hand, the $\varepsilon - \delta$ condition clearly implies that $\nu \ll \mu$. On the other hand, if the $\varepsilon - \delta$ condition is not satisfied, then there exists an $\varepsilon_0 > 0$ such that for any $\delta > 0$, we can find a set $A \in \mathscr{A}$ with $\mu(A) < \delta$ and $|\nu|(A) \geq \varepsilon_0$. As a result, there exists an $A_k \in \mathscr{A}$ satisfying $\mu(A_k) < 2^{-k}$ and $|\nu|(A_k) \geq \varepsilon_0$ for all $k \in \mathbb{N}$. Let $A = \overline{\lim}_{k \to \infty} A_k$, then for all $k \in \mathbb{N}$,

$$\mu(A) \leq \mu \left(\bigcup_{n=k}^{\infty} A_n \right) \leq \sum_{n=k}^{\infty} 2^{-n} = 2^{-k+1},$$

thus $\mu(A) = 0$. Since ν is finite,

$$|\nu|(A) = |\nu| \left(\overline{\lim_{k \to \infty}} A_k \right) \geq \overline{\lim_{k \to \infty}} |\nu|(A_k) \geq \varepsilon_0,$$

which implies ν is not absolutely continuous. Thereby, the necessity is obtained. $\square$

Example 1.4.3. Suppose μ is a usual Lebesgue measure on $\mathbb{R}^n$. Let ν satisfy that $\nu = 0$ on all null sets with respect to φ and $\nu = \infty$ on all positive sets with respect to φ, then ν is a measure with $\nu \ll \mu$ but clearly does not meet the condition of Theorem 1.4.6. This indicates that the necessity of Theorem 1.4.6 does not hold without the condition of ν being finite.

Theorem 1.4.7. *Let $(X, \mathscr{A}, \mu)$ be a measure space and ν be a signed measure on $(X, \mathscr{A})$. Then $\nu \perp \mu$ if and only if for any $\varepsilon > 0$, there exists a set $E \in \mathscr{A}$ with $\mu(E) < \varepsilon$ and $|\nu|(\complement E) < \varepsilon$.*

Proof.   On the one hand, the necessity is obvious. On the other hand, if the condition given holds, then there exists an $E_k \in \mathscr{A}$ such that

$$\mu(E_k) < 2^{-k}, \quad |\nu|(\complement E_k) < 2^{-k}.$$

Let $Z = \overline{\lim}_{k \to \infty} E_k$, then $\mu(Z) = 0$ and

$$\complement Z = \bigcup_{n=1}^{\infty} \bigcap_{k=n}^{\infty} \complement E_k,$$

$$|\nu|(\complement Z) = \lim_{n \to \infty} |\nu| \left(\bigcap_{k=n}^{\infty} \complement E_k \right) \leq \overline{\lim}_{n \to \infty} |\nu|(\complement E_n) = 0.$$

Thus, $\nu \perp \mu$. $\square$

Exercise 1.4

1. Let μ and ν be signed measures on a measurable space $(X, \mathscr{A})$. Prove that
   (1) $\mu \ll \nu$ and $\mu \perp \nu$ if and only if $\mu = 0$;
   (2) $\mu \ll \nu \Leftrightarrow \mu^+ \ll \nu$, $\mu^- \ll \nu \Leftrightarrow |\mu| \ll \nu$;
   (3) both $\{\varphi : \varphi$ is a finite signed measure on $(X, \mathscr{A})$ and $\varphi \ll \nu\}$ and $\{\varphi : \varphi$ is a finite signed measure on $(X, \mathscr{A})$ and $\varphi \perp \nu\}$ are real linear spaces.

2. Let $(X, \mathscr{A}, \mu)$ be a finite measure space and $\{\nu_n\}_{n=1}^{\infty}$ be a sequence of finite measures with $\nu_n \ll \mu$. Show that if $\lim_{n \to \infty} \nu_n(E) = \nu(E)$ are all finite for every $E \in \mathscr{A}$, then
   (1) ν_n are uniformly absolutely continuous with respect to μ, that is,

$$\lim_{\mu(E) \to 0} \nu_n(E) = 0$$

   holds uniformly with respect to $n \in \mathbb{N}$;
   (2) ν is a finite measure.

Hint: Prove that ν_n is continuous on the metric space $(\mathscr{A}, d)$, which is defined as in Exercise 3 in Section 1.1. For any given $\varepsilon > 0$, consider the family of sets

$$\mathscr{A}_{m,n} = \{E \in \mathscr{A} : |\nu_n(E) - \nu_m(E)| \leq \varepsilon\}, \quad m, n \in \mathbb{N}$$

and

$$\mathscr{A}_p = \bigcap_{m,n \geq p} \mathscr{A}_{m,n}, \quad n \in \mathbb{N},$$

then apply the Baire category theorem.

3. Let $(X, \mathscr{A})$ be a measurable space and $\{\nu_n\}_{n=1}^{\infty}$ be a sequence of nonzero finite signed measures satisfying $\lim_{n \to \infty} \nu_n(E) = \nu(E)$ exists and is finite for every $E \in \mathscr{A}$. Prove that ν is a finite signed measure.
   Hint: Let $\mu(E) = \sum_{n=1}^{\infty} \frac{\nu_n(E)}{|\nu_n|(X)} 2^{-n}$ and apply Exercise 2. The conclusion of Exercise 2 can be extended to the finite signed measure.

4. For $n \in \mathbb{N}$, suppose $(X, \mathscr{A}, \mu_n)$ is a measure space. Let $\mu = \sum_{n=1}^{\infty} \mu_n$, then μ is a measure on $(X, \mathscr{A})$.

5. Give an example to illustrate the existence of a signed measure ν and a measurable set E satisfying $\nu(E) = 0$, but E is not null.

6. Let ν be a signed measure on a measurable space $(X, \mathscr{A})$. For all $E \in \mathscr{A}$, show that

$$|\nu|(E) = \sup \left\{ \sum_{k=1}^{n} |\nu(E_k)| : E = \bigcup_{k=1}^{n} E_k, E_k \in \mathscr{A}, E_k \cap E_j = \emptyset \right.$$

$$\left. \text{when } k \neq j; n \in \mathbb{N} \right\}.$$

7. Let μ be a finite signed measure on a measurable space $(X, \mathscr{A})$ and $E \in \mathscr{A}$. Prove that

$$|\mu|(E) = \sup \left\{ \left| \int_E f \, d\mu \right| : f \in L(X, \mathscr{A}, |\mu|), |f| \leq 1 \right\}.$$

1.5 The Radon–Nikodym Theorem

Let $(X, \mathscr{A})$ be a measurable space and φ be a signed measure on $\mathscr{A}$. According to Section 1.4, we have $\varphi = \varphi^+ - \varphi^-$, where φ^+ and φ^-

are measures and at least one of them is finite. Obviously, according to Definition 1.1.4, the proposition, φ is finite, is equivalent to both φ^+ and φ^- being finite; the proposition, φ is σ-finite, is equivalent to both φ^+ and φ^- being σ-finite.

Lemma 1.5.1. *Let μ and ν be finite measures on the measurable space $(X, \mathscr{A})$. If $\nu \ll \mu$, then there exists an $\mathscr{A}$-measurable function g such that $g(X) \subset [0,1)$ and for any $f \in L^2(X, \mathscr{A}, \mu + \nu)$,*

$$\int_X f d\nu = \int_X fg d\nu + \int_X fg d\mu.$$

Proof. For $f \in L^2(X, \mathscr{A}, \mu + \nu)$, define

$$L(f) = \int_X f d\nu,$$

then

$$|L(f)| \leq \left(\int_X |f|^2 d\nu \right)^{\frac{1}{2}} \sqrt{\nu(X)} \leq \sqrt{\nu(X)} \, \|f\|_{L^2(X, \mathscr{A}, \mu + \nu)}.$$

It is thus clear that L is a bounded linear functional on $L^2(X, \mathscr{A}, \mu + \nu)$. According to the representation theorem, that is, Corollary 1.3.12, there exists an $h \in L^2(X, \mathscr{A}, \mu + \nu)$ such that for all $f \in L^2(X, \mathscr{A}, \mu + \nu)$,

$$L(f) = \int_X f h d(\mu + \nu).$$

Let $A = \{x \in X : h(x) < 0\}$, then

$$0 \leq L(\chi_A) = \nu(A) = \int_A h d(\mu + \nu).$$

Hence, one can see that $(\mu + \nu)(A) = 0$. Set $E = \{x \in X : h(x) \geq 1\}$. Due to $\chi_E \in L^2(X, \mathscr{A}, \mu + \nu)$,

$$0 \leq \mu(E) \leq \int_X \chi_E h d\mu = L(\chi_E) - \int_X \chi_E h d\nu = \int_X \chi_E (1 - h) d\nu \leq 0.$$

Thus, $\mu(E) = 0$. Combined with $\nu \ll \mu$, we have $\nu(E) = 0$. Let $g = h\chi_{\complement(E \cup A)}$, then $g(X) \subset [0,1)$ and for all $f \in L^2(X, \mathscr{A}, \mu + \nu)$,

$$\int_X f d\nu = \int_X f h d(\mu + \nu) = \int_X fg d(\mu + \nu).$$

$\square$

Theorem 1.5.2 (Radon–Nikodym's theorem). *Let μ and ν be finite measures on a measurable space $(X, \mathscr{A})$. If $\nu \ll \mu$, then there exists a nonnegative function $f_0 \in L(X, \mathscr{A}, \mu)$ such that for all nonnegative measurable function f,*

$$\int_X f d\nu = \int_X f f_0 d\mu,$$

especially for any $A \in \mathscr{A}$,

$$\nu(A) = \int_A f_0 d\mu.$$

Proof. Suppose g is the function described in Lemma 1.5.1. If f is a nonnegative bounded measurable function, then

$$(1 + g + \cdots + g^{n-1})f \in L^2(X, \mathscr{A}, \mu + \nu)$$

for all $n \in \mathbb{N}$. Lemma 1.5.1 yields

$$\int_X (1 + g + \cdots + g^{n-1})f(1 - g)d\nu = \int_X (1 + g + \cdots + g^{n-1})fg d\mu,$$

i.e.

$$\int_X (1 - g^n)f d\nu = \int_X \frac{1 - g^n}{1 - g} g f d\mu.$$

Let $n \to \infty$, using the Levi theorem, we obtain

$$\int_X f d\nu = \int_X \frac{g}{1 - g} f d\mu.$$

Substituting $f = 1$ yields $\frac{g}{1-g} \in L(X, \mathscr{A}, \mu)$. Finally, letting $f_0 = \frac{g}{1-g}$ gives

$$\int_X f d\nu = \int_X f f_0 d\mu.$$

$\square$

Theorem 1.5.3 (Radon–Nikodym's theorem). *Let μ and ν be σ-finite measures on a measurable space $(X, \mathscr{A})$. If $\nu \ll \mu$, then there exists a unique (in the sense of μ-a.e.) nonnegative finite $\mathscr{A}$-measurable function f_0 such that*

(1) $\int_X f d\nu = \int_X f f_0 d\mu$ for all nonnegative measurable functions f and all $f \in L(X, \mathscr{A}, \nu)$;

(2) $\nu(A) = \int_A f_0 d\mu$ for all $A \in \mathscr{A}$.

Proof. Since μ and ν are σ-finite, X can be decomposed into a pairwise disjoint measurable sequence of sets $\{E_n\}_{n=1}^{\infty}$ satisfying $\mu(E_n) < \infty$ and $\nu(E_n) < \infty$. For $n \in \mathbb{N}$ and any $A \in \mathscr{A}$, define

$$\mu_n(A) = \mu(A \cap E_n) \quad \text{and} \quad \nu_n(A) = \nu(A \cap E_n),$$

then μ_n and ν_n are finite measures on $(X, \mathscr{A})$ and $\nu_n \ll \mu_n$. According to Theorem 1.5.2, there exists a nonnegative finite measurable function f_n such that, for any nonnegative measurable function f,

$$\int_X f d\nu_n = \int_X f f_n d\mu_n.$$

Let $f_0 = \sum_{n=1}^{\infty} f_n \chi_{E_n}$ which is a nonnegative finite measurable function. The Levi theorem yields that, for a nonnegative measurable function f,

$$\int_X f d\nu = \sum_{n=1}^{\infty} \int_X f \chi_{E_n} d\nu = \sum_{n=1}^{\infty} \int_X f d\nu_n = \sum_{n=1}^{\infty} \int_X f f_n d\mu_n$$

$$= \sum_{n=1}^{\infty} \int_X f f_n \chi_{E_n} d\mu = \int_X f f_0 d\mu.$$

Thus, (1) can be inferred and (2) is obtained by substituting $f = \chi_A$ in the above equation. The uniqueness of f_0 is clear. $\qquad \square$

Definition 1.5.1. The function f_0 in Theorem 1.5.3 is called the Radon–Nikodym derivative of ν with respect to μ, denoted by $\frac{d\nu}{d\mu}$, which is also often represented by the notation

$$d\nu = f_0 d\mu \quad \text{and} \quad \nu = f_0 \mu.$$

Remark. The name of the Radon–Nikodym theorem and the Radon–Nikodym derivative in some books are prefixed with Lebesgue.

For a signed measure φ, as defined in Section 1.4,

$$\int_X f d\varphi = \int_X f d\varphi^+ - \int_X f d\varphi^- \quad \text{for all } f \in L(X, \mathscr{A}, |\varphi|).$$

The Radon–Nikodym theorem and the Hahn decomposition theorem can be used to derive corresponding results about signed measure.

Theorem 1.5.4. *Let $(X, \mathscr{A}, \mu)$ be a σ-finite measure space and φ be a σ-finite signed measure on $(X, \mathscr{A})$. If $\varphi \ll \mu$, then there exists a unique, in the sense of μ-a.e., measurable function f_0 which takes finite values only such that*

(1) $\int_X f d\varphi = \int_X f f_0 d\mu$ *for all $f \in L(X, \mathscr{A}, |\varphi|)$;*
(2) $\varphi(A) = \int_A f_0 d\mu$ *for all $A \in \mathscr{A}$;*
(3) $|\varphi|(A) = \int_A |f_0| d\mu$ *for all $A \in \mathscr{A}$;*

and f_0 is unique in the sense of μ-a.e.

Proof. According to the Hahn decomposition theorem and Theorem 1.4.4, it is known that there exists a $P \in \mathscr{A}$ such that, for all $E \in \mathscr{A}$,

$$\varphi^+(E) = \varphi(E \cap P), \quad \varphi^-(E) = -\varphi(E \cap \complement P).$$

Define $\mathscr{A} \cap P = \{E \cap P : E \in \mathscr{A}\}$ and $\mathscr{A} \cap \complement P = \{E \cap \complement P : E \in \mathscr{A}\}$. Using Theorem 1.5.3 for $\varphi^+ \ll \mu$ and $\varphi^- \ll \mu$ on two σ-finite measure spaces $(P, \mathscr{A} \cap P, \mu)$ and $(\complement P, \mathscr{A} \cap \complement P, \mu)$, respectively, it can be concluded that there exist nonnegative finite measurable functions f_0^+ on P and f_0^- on $\complement P$ satisfying

$$\int_P f d\varphi^+ = \int_P f f_0^+ d\mu \quad \text{for any } f \in L(P, \mathscr{A} \cap P, \varphi^+) \qquad \text{(a)}$$

and

$$\int_{\complement P} f d\varphi^- = \int_{\complement P} f f_0^- d\mu \quad \text{for any } f \in L(\complement P, \mathscr{A} \cap \complement P, \varphi^-). \qquad \text{(b)}$$

Thus, for all $f \in L(X, \mathscr{A}, |\varphi|)$,

$$\int_X f d\varphi = \int_X f d\varphi^+ - \int_X f d\varphi^- = \int_P f d\varphi^+ - \int_{\complement P} f d\varphi^-.$$

Derived from (a) and (b),

$$\int_X f d\varphi = \int_P f f_0^+ d\mu - \int_{\complement P} f f_0^- d\mu. \tag{c}$$

Define

$$f_0(x) = \begin{cases} f_0^+(x) & \text{if } x \in P, \\ -f_0^-(x) & \text{if } x \in \complement P. \end{cases}$$

From (c),

$$\int_X f d\varphi = \int_X f f_0 d\mu,$$

which completes the proof of (1). Thus, (2) is obtained. (3) is also derived from (a) and (b). $\square$

Similar to Definition 1.5.1, f_0 in Theorem 1.5.4 is called the R–N derivative of φ with respect to μ, denoted by $f_0 = \frac{d\varphi}{d\mu}$ or $d\varphi = f_0 d\mu$, $\varphi = f_0 \cdot \mu$.

Corollary 1.5.5. *Let φ be a σ-finite signed measure on $(X, \mathscr{A})$. Then there exists a measurable function f_0 satisfying*

(1) $\int_X f d\varphi = \int_X f f_0 d|\varphi|$ *for all* $f \in L(X, \mathscr{A}, |\varphi|)$;
(2) $\varphi(A) = \int_A f_0 d|\varphi|$ *for all* $A \in \mathscr{A}$;
(3) $|f_0| = 1$.

Proof. It is obvious that $\varphi \ll |\varphi|$. Theorem 1.5.4 yields that there exists an $f_0 \in L(X, \mathscr{A}, |\varphi|)$ such that (1) and (2) hold.

Let $A = \{x : |f_0(x)| < 1\}, B = \{x : |f_0(x)| > 1\}$. Applying Theorem 1.5.4(3), we obtain

$$|\varphi|(A) = \int_A |f_0| \, d|\varphi|, \quad |\varphi|(B) = \int_B |f_0| \, d|\varphi|,$$

which indicates that

$$\int_A (1 - |f_0|) \, d|\varphi| = 0 = \int_B (|f_0| - 1) \, d|\varphi|.$$

Thereby,

$$|\varphi|(A) = |\varphi|(B) = 0.$$

If $f_0 = 1$ on $A \cup B$, then (3) holds. $\qquad\square$

Remark. We can directly prove from the definition that, for any $f \in L(X, \mathscr{A}, |\varphi|)$,

$$\left| \int f \, d\varphi \right| \leq \int |f| d|\varphi|.$$

According to Corollary 1.5.5, the above inequality can be derived as follows:

$$\left| \int f \, d\varphi \right| = \left| \int f f_0 d|\varphi| \right| \leq \int |f f_0| d|\varphi| = \int |f| d|\varphi|.$$

Theorem 1.5.6 (Lebesgue's decomposition theorem). *Let $(X, \mathscr{A}, \mu)$ be a σ-finite measure space and ν be a σ-finite singed measure on $(X, \mathscr{A})$. Then ν can be uniquely decomposed into the sum of two σ-finite signed measures: $\nu = \nu_1 + \nu_2$ satisfying $\nu_1 \ll \mu$, $\nu_2 \perp \mu$.*

Proof. Assume ν is a measure. According to Theorem 1.5.3, since $\nu \ll \mu + \nu$, there exists a nonnegative measurable function f_0 such that

$$\int_X f d\nu = \int_X f f_0 d(\mu + \nu) \quad \text{for all } f \in L(X, \mathscr{A}, \nu).$$

If there exists an $\varepsilon > 0$ such that $F := \{x \in X : f_0(x) \geq 1 + \varepsilon\}$ satisfying $(\mu + \nu)(F) > 0$, then there exists an $A \subset F$ with $0 < (\mu + \nu)(A) < \infty$ on account of that $\mu + \nu$ is σ-finite. Therefore,

$$\int_X \chi_A d\nu = \int_X \chi_A f_0 d(\mu + \nu),$$

$$\nu(A) = \int_A f_0 d(\mu + \nu) \ge (1 + \varepsilon)(\mu + \nu)(A),$$

$$(1 + \varepsilon)\mu(A) + \varepsilon\nu(A) \le 0.$$

Hence, $\nu(A) = \mu(A) = 0$, which contradicts that $(\mu + \nu)(A) > 0$ and shows

$$(\mu + \nu)(\{x : f_0(x)\} > 1) = 0.$$

Let $f_1 = \min(f_0, 1)$, then $0 \le f_1 \le 1$ and

$$\int_X f d\nu = \int_X f f_1 d(\mu + \nu) \quad \text{for all } f \in L(X, \mathscr{A}, \nu).$$

Let $B = \{x : f_1(x) = 1\}$ and $C \subset B$ with $C \in \mathscr{A}$. As long as

$$\mu(C) < \infty, \quad \nu(C) < \infty,$$

there exists

$$\nu(C) = \int_X \chi_C d\nu = \int_C f_1 d(\mu + \nu) = \mu(C) + \nu(C),$$

which shows $\mu(C) = 0$. Then $\mu(B) = 0$ due to the σ-finiteness of μ and ν.

Define ν_1 and ν_2 as follows:

$$\nu_1(A) = \nu(A \cap \complement B), \quad \nu_2(A) = \nu(A \cap B) \quad \text{for any } A \in \mathscr{A}.$$

Then ν_1 and ν_2 are both σ-finite measures and $\nu = \nu_1 + \nu_2$. It is clear that $\nu_2(\complement B) = 0$. Hence, $\nu_2 \perp \mu$.

Assume $A \in \mathscr{A}$ with $\mu(A) = 0$. If $\nu(A) < \infty$, then

$$\nu_1(A) = \nu(A \cap \complement B) = \int_{A \cap \complement B} f_1 d(\mu + \nu) = \int_{A \setminus B} f_1 d\nu,$$

thus

$$\int_{A \setminus B} (1 - f_1) d\nu = 0.$$

Since $1 - f_1 > 0$ on $A \setminus B$, $\nu(A \setminus B) = 0$, that is, $\nu_1(A) = 0$. If $\nu(A) = \infty$, it can still be inferred from the known results that $\nu_1(A) = 0$ since ν, ν_1 are σ-finite. Thereby, $\nu_1 \ll \mu$.

Suppose ν is a signed measure, using the obtained results for ν^+ and ν^-, it can be inferred that $\nu^+ = \nu_1^+ + \nu_2^+, \nu^- = \nu_1^- + \nu_2^-$ with

$$\nu_1^+ \ll \mu, \quad \nu_1^- \ll \mu, \quad \nu_2^+ \perp \mu, \quad \nu_2^- \perp \mu.$$

At least one of ν^+ and ν^- is a finite measure, as a result, $\nu_1^+ - \nu_1^-$ and $\nu_2^+ - \nu_2^-$ are both σ-finite measures, denoted by ν_1 and ν_2, respectively. Then $\nu = \nu_1 + \nu_2$ is the desired decomposition.

The uniqueness of Lebesgue decomposition is evident. □

Here, we need to point out that when considering the uniqueness problem, it is not enough to deduce $\nu_2 = \nu_2'$ from $\nu_1 + \nu_2 = \nu_1' + \nu_2'$ and $\nu_1 = \nu_1'$ as we are dealing with generalized real numbers.

As an application of the Radon–Nikodym theorem, the dual space of $L^p(X, \mathscr{A}, \mu)$ $(1 \leq p < \infty)$ is worth discussing.

Theorem 1.5.7. *Let* $1 \leq p < \infty, \frac{1}{p} + \frac{1}{p'} = 1$ *and* $\frac{1}{\infty} = 0$. *Suppose* $(X, \mathscr{A}, \mu)$ *is a* σ-*finite measure space, then, for any* $\iota \in [L^p(X, \mathscr{A}, \mu)]^*$, *there exists a unique* $g \in L^{p'}(X, \mathscr{A}, \mu)$ *such that*

$$\iota(f) = \int_X fg d\mu \quad and \quad \|g\|_{p'} = \|\iota\| \quad for \ all \ f \in L^p(X, \mathscr{A}, \mu).$$

Proof. Consider $\mu(X) < \infty$ first. Define $\varphi : \varphi(A) = \iota(\chi_A)$ for all $A \in \mathscr{A}$. Then

$$|\varphi(A)| \leq \|\iota\| \quad \|\chi_A\|_p \leq \|\iota\| \ [\mu(A)]^{\frac{1}{p}} \leq \|\iota\| \ [\mu(X)]^{\frac{1}{p}} < \infty.$$

Let $A = \cup_{k=1}^{\infty} A_k$ with $A_k \in \mathscr{A}$ being pairwise disjoint. Since ι is additive,

$$\varphi(A) = \varphi\left(\bigcup_{k=1}^{m} A_k\right) + \varphi\left(\bigcup_{k=m+1}^{\infty} A_k\right).$$

Note that

$$\mu\left(\bigcup_{k=m+1}^{\infty} A_k\right) = \sum_{k=m+1}^{\infty} \mu(A_k) \to 0 \ (m \to \infty),$$

which yields

$$\varphi\left(\bigcup_{k=m+1}^{\infty} A_k\right) \to 0 \quad (m \to \infty).$$

Thus,

$$\varphi(A) = \sum_{k=1}^{\infty} \varphi(A_k),$$

which demonstrates convincingly that φ is a finite signed measure and $\varphi \ll \mu$.

According to Theorem 1.5.4, there exists a $g \in L(X, \mathscr{A}, \mu)$ such that for all $A \in \mathscr{A}$,

$$\varphi(A) \int_A g d\mu = \int_X \chi_A g d\mu = \iota(\chi_A).$$

Thereby, for any simple function f,

$$\iota(f) = \int_X f g d\mu$$

holds.

When $1 < p < \infty$, take nonnegative increasing simple functions $h_k (k \in \mathbb{N})$ converging to $|g|$ and let

$$g_k = h_k^{p'-1} \operatorname{sgn} g \quad (k \in \mathbb{N}),$$

then g_k are also simple functions and

$$\iota(g_k) = \int_X h_k^{p'-1} |g| d\mu \le \|\iota\| \quad \|g_k\|_p = \|\iota\| \left(\int_X h_k^{p'} d\mu \right)^{\frac{1}{p}}.$$

Combined with $h_k \le |g|$,

$$\int_X h_k^{p'} d\mu \le \iota(g_k).$$

Thus,

$$\|h_k\|_{p'} \le \|\iota\|.$$

Let $k \to \infty$, the Levi theorem yields

$$\|g\|_{p'} \le \|\iota\|.$$

When $p = 1$, for any $A \in \mathscr{A}$, let $h_A = \chi_A \operatorname{sgn} g$, then

$$\iota(h_A) = \int_A |g| d\mu \le \|\iota\| \quad \|h_A\|_1 \le \|\iota\| \, \mu(A).$$

Hence, $g \in L^{p'} = L^{\infty}$ and $\|g\|_{\infty} \leq \|\iota\|$. In short, $g \in L^{p'}, \|g\|_{p'} \leq \|\iota\|$ and for a simple function f,

$$\iota(f) = \int_X fg d\mu.$$

Let $f \in L^p(= L^p(X, \mathscr{A}, \mu))$, there exist simple functions $f_k(k \in \mathbb{N})$ satisfying $\|f_k - f\|_p \to 0 (k \to \infty)$. Thus,

$$\iota(f) = \iota(f - f_k) + \iota(f_k) = \iota(f - f_k) + \int_X (f_k - f)g d\mu + \int_X fg d\mu.$$

Note that $\iota(f - f_k) \to 0$ and $\int_X (f_k - f)g d\mu \to 0$ as $k \to \infty$, then

$$\iota(f) = \int_X fg d\mu.$$

By Hölder's inequality,

$$|\iota(f)| \leq \|f\|_p \|g\|_{p'}.$$

Therefore, $\|\iota\| \leq \|g\|_{p'}$. All results above indicate $\|\iota\| = \|g\|_{p'}$.

The uniqueness of g is evident.

Then we consider the case $\mu(X) = \infty$. Bearing that μ is σ-finite in mind, by decomposing X into a countable union of finite measures and transitioning the proved results to the limit case, the conclusions of the theorem can be obtained. □

In our argument system, the representation theorem of bounded linear functional on L^2 space is used to prove the R–N theorem. As a result, logically, the proof of Theorem 1.5.7 is only valid for $p \neq 2$. According to Theorem 1.5.7, we write $(L^p)^* = L^{p'}(1 \leq p < \infty)$ in the sense of isometric isomorphism.

Exercise 1.5

1. Let $(X, \mathscr{A}, \mu)$ be a measure space and define ν:

$$\nu(A) = 0 \quad \text{if } \mu(A) = 0,$$
$$\nu(A) = \infty \quad \text{if } \mu(A) > 0.$$

Prove that $(X, \mathscr{A}, \nu)$ is a measure space and $\nu \ll \mu$. Construct f_0 such that Theorem 1.5.3 holds for μ and ν.

2. Prove the chain rule: let μ_0, μ_1, μ_2 be σ-finite measures on $(X, \mathscr{A})$ with $\mu_2 \ll \mu_1, \mu_1 \ll \mu_0$, then $\mu_2 \ll \mu_0$ and

$$\frac{d\mu_2}{d\mu_0} = \frac{d\mu_2}{d\mu_1} \cdot \frac{d\mu_1}{d\mu_0} \quad \mu_0\text{-a.e.}$$

3. Let μ, ν be σ-finite measures on $(X, \mathscr{A})$ with $\nu \ll \mu$ and $\mu \ll \nu$. Show that

$$\frac{d\nu}{d\mu} \neq 0 \text{ a.e.,} \quad \text{and} \quad \frac{d\mu}{d\nu} = \left(\frac{d\nu}{d\mu}\right)^{-1} \text{ a.e.,}$$

where μ-a.e. and ν-a.e. are the same.

4. Let $\mathscr{A} = \{A \subset \mathbb{R} : \min(\text{card}(A), \text{card}(\complement A)) \leq \aleph_0\}$. Prove that $\mathscr{A}$ is a σ-algebra. Define μ and ν as follows: for any $A \in \mathscr{A}$,

$$\mu(A) = \begin{cases} \text{card}(A) & \text{if card}(A) < \aleph_0, \\ \infty & \text{if card}(A) \geq \aleph_0, \end{cases}$$

$$\nu(A) = \begin{cases} 0 & \text{if card}(A) \leq \aleph_0, \\ \infty & \text{if card}(A) = \aleph_1. \end{cases}$$

Then μ and ν are measures on $(\mathbb{R}, \mathscr{A})$. Please prove $\nu \ll \mu$, but the R–N theorem does not hold in this case.

5. Let $(X, \mathscr{A}, \mu)$ be a σ-finite measure space, $1 \leq p \leq \infty$ and f be measurable. Set

$$\mathscr{F} = \left\{ g : \|g\|_{p'} \leq 1, \mu(\{x \in X : g(x) \neq 0\}) < \infty \right.$$

$$\text{and} \left. \int_X fg d\mu \text{ exists} \right\},$$

where p' is the conjugate exponent to p, that is, $\frac{1}{p} + \frac{1}{p'} = 1$ and $\frac{1}{\infty} = 0$, show that

$$\|f\|_p = \sup \left\{ \int_X fg d\mu : g \in \mathscr{F} \right\}.$$

6. Prove Theorem 1.5.7 in detail when $\mu(X) = \infty$.

1.6   Outer Measure

This section is divided into five parts, of which the fifth part can be omitted.

1.6.1   *Induce a Measure from an Outer Measure*

Definition 1.6.1. Let X be a set, $\mathscr{M} = 2^X$. An outer measure on X is a generalized real-valued function μ^* that satisfies the following:

(1) $\mu^*(A) \geq 0$ for all $A \in \mathscr{M}$, $\mu^*(\emptyset) = 0$,
(2) $A \subset B \Rightarrow \mu^*(A) \leq \mu^*(B)$,
(3) $E = \cup_{i=1}^{\infty} E_i \Rightarrow \mu^*(E) \leq \sum_{i=1}^{\infty} \mu^*(E_i)$,

where (3) is called the countable subadditivity.

An outer measure μ^* is finite if $\mu^*(X) < \infty$ or σ-finite if there exists an E_i satisfying $\mu^*(E_i) < \infty$ for every $i \in \mathbb{N}$ and $X = \cup_{i=1}^{\infty} E_i$.

Example 1.6.1. Lebesgue outer measures on $\mathbb{R}^n$.
   A set of the form $I = \{x \in \mathbb{R}^n : a_i < x_i < a_i + l_i, a_i \in \mathbb{R}, l_i \in (0, \infty), i = 1, \ldots, n\}$ is called a rectangle of $\mathbb{R}^n$. For convenience, the empty set $\emptyset$ is also called a rectangle of side length 0. Stipulate that the volume of I is $l_1 l_2 \cdots l_n$, which is denoted by $|I|$. Define

$$m^*(E) = \inf \left\{ \sum_{k=1}^{\infty} |I_k| : I_k \text{ are rectangles}, \bigcup_{k=1}^{\infty} I_k \supset E \right\}$$
$$\text{for any } E \subset \mathbb{R}^n,$$

which is the Lebesgue outer measure that has been discussed in detail in theory of functions of real variable.

Definition 1.6.2. If μ^* is an outer measure on X, then a set $E \subset X$ is said to be a μ^*-measurable set, abbreviated as a measurable set, provided the following condition holds: For all $A \subset X$,

$$(C) \quad \mu^*(A) = \mu^*(A \cap E) + \mu^*(A \cap \complement E).$$

Remark. The condition (C) is called the mathematician Carathéodory.

Theorem 1.6.1. *If μ^* is an outer measure on X, then the collection $\mathscr{A}$ of all μ^*-measurable sets is a σ-algebra, and the restriction μ of μ^* to $\mathscr{A}$ is a complete measure, that is, $(X, \mathscr{A}, \mu)$ is a complete measure space. μ is called the measure induced by μ^*, and $(X, \mathscr{A}, \mu)$ is said to be induced by μ^*.*

This theorem can be proved by the following steps:

(1) Let $A_1, A_2 \in \mathscr{A}$, prove that $A_1 \cap A_2 \in \mathscr{A}$.
(2) Show the following lemma by induction:

> **Lemma 1.6.2.** *Let $A_k \in \mathscr{A}, A_k \cap A_j = \emptyset (k \neq j)$ and $A = \cup_{k=1}^{\infty} A_k$, then*
>
> $$\mu^*(E) = \sum_{k=1}^{\infty} \mu^*(E \cap A_k) + \mu^*(E \cap \complement A) \quad \textit{for all } E \subset X.$$
>
> Replacing E with $E \cap A$ in the above equation,
>
> $$\mu^*(E \cap A) = \sum_{k=1}^{\infty} \mu^*(E \cap A_k)$$
>
> is obtained, thus $A \in \mathscr{A}$.

(3) Prove that $(X, \mathscr{A}, \mu)$ is a complete measure space applying the conclusions of (1) and (2).

Note that the measure m, induced by m^* in Example 1.6.1, is just the usual Lebesgue measure, which has been studied in detail in theory of functions of real variable. In order to further study the outer measure and its induced measure on $\mathbb{R}^n$, especially to introduce the Lebesgue–Stielties measure on $\mathbb{R}^n$, we first discuss the outer measure related to metric in the metric space.

Definition 1.6.3. *Let (X, d) be a metric space. An outer measure μ^* on X is called a metric outer measure on X or (X, d) provided, for any $E_1, E_2 \subset X$,*

$$\mu^*(E_1 \cup E_2) = \mu^*(E_1) + \mu^*(E_2)$$

when $\rho(E_1, E_2) := \inf\{d(x_1, x_2) : x_1 \in E_1, x_2 \in E_2\} > 0$.

Definition 1.6.4. Let (X, d) be a metric space. The Borel algebra is the σ-algebra generated by the family of the open sets in X, i.e. the smallest σ-algebra containing all open sets in X, which is denoted by $\mathscr{B}$.

It should be noted that there is no unified regulation on the definition of the Borel algebra. For example, in some books, the Borel algebra is defined as the σ-algebra generated by all compact sets, and the σ-algebra generated by all open sets is called the weak Borel algebra.

Theorem 1.6.3. *Let μ^* be a metric outer measure on a metric space (X, d) and $\mathscr{A}$ be the collection of all μ^*-measurable sets, then $\mathscr{A} \supset \mathscr{B}$.*

Proof. It will be proved that an open set is μ^*-measurable. Let G be any open set with $G \neq X$. For a subset $A \subset G$, set

$$A_k = \left\{ x \in A : \rho(\{x\}, \complement G) \geq \frac{1}{k} \right\}, \quad k \in \mathbb{N}.$$

It is clear that $A_k \nearrow A$. Let $B_1 = A_1, B_{k+1} = A_{k+1} \backslash A_k, k \in \mathbb{N}$, then A can be decomposed into pairwise disjoint sets:

$$A = A_k \cup \left(\bigcup_{j=k+1}^{\infty} B_j \right) \quad \text{for all } k \in \mathbb{N}.$$

By countable subadditivity,

$$\mu^*(A) \leq \mu^*(A_k) + \sum_{j=k+1}^{\infty} \mu^*(B_j).$$

When $\sum_{j=1}^{\infty} \mu^*(B_j) < \infty$, let $k \to \infty$,

$$\mu^*(A) \leq \lim_{k \to \infty} \mu^*(A_k).$$

Bearing that $A \supset A_k$ and μ^* is monotonic in mind,

$$\mu^*(A) = \lim_{k \to \infty} \mu^*(A_k).$$

When $\sum_{j=1}^{\infty} \mu^*(B_j) = \infty$, for $k \geq 3$,

$$A_k \supset \bigcup_{0 \leq 2j < k} B_{k-2j} \quad \text{or} \quad A_k \supset \bigcup_{0 \leq 2j < k-1} B_{k-2j-1}.$$

Note that $\rho(A_j, B_{j+2}) > 0$ and μ^* is a metric outer measure, thus

$$\mu^*(A_k) \geq \sum_{0 \leq 2j < k} \mu^*(B_{k-2j}) \quad \text{or} \quad \mu^*(A_k) \geq \sum_{0 \leq 2j < k-1} \mu^*(B_{k-2j-1}).$$

Therefore,

$$\mu^*(A) \geq \mu^*(A_k) \geq \frac{1}{2} \sum_{j=1}^{k} \mu^*(B_j) \to \infty \ (k \to \infty),$$

which implies that $\mu^*(A) = \lim_{k\to\infty} \mu^*(A_k)$.

For any $E \subset X$, let $E \cap G = A, E \cap \complement G = B$ and $A_k = \{x \in A : \rho(\{x\}, \complement G) \geq \frac{1}{k}\}, k \in \mathbb{N}$, then $\rho(A_k, B) > 0$,

$$\mu^*(E) \geq \mu^*(A_k \cup B) = \mu^*(A_k) + \mu^*(B).$$

Let $k \to \infty$, the previous results yield

$$\mu^*(E) \geq \mu^*(A) + \mu^*(B),$$

which indicates that the open set G meets the condition (C). Hence, $G \in \mathscr{A}$.

It is immediate to verify that $\mathscr{B} \subset \mathscr{A}$ since $\mathscr{B}$ is the smallest σ-algebra containing all open sets. $\qquad\square$

Definition 1.6.5. Let μ^* be an outer measure on X and $(X, \mathscr{A}, \mu)$ be the measure space induced by μ^*. The outer measure μ^* is called regular provided for any $E \subset X$, there exists an $A \in \mathscr{A}$ such that

$$E \subset A \text{ and } \mu^*(E) = \mu(A).$$

1.6.2   The Lebesgue–Stieltjes Outer Measures on $\mathbb{R}$

Let f be a real-valued monotonic increasing function. For a finite interval $(a, b] \subset \mathbb{R}$, define

$$\lambda_f((a, b]) = f(b) - f(a).$$

Specify that $(a, b] = \emptyset$ when $a = b$. For $A \subset \mathbb{R}$, define

$$\Lambda_f^*(A) = \inf \left\{ \sum_{k=1}^{\infty} \lambda_f((a_k, b_k]) : \bigcup_{k=1}^{\infty} (a_k, b_k] \supset A \right\},$$

which indicates $\Lambda_f^*(\emptyset) = 0$. In the light of Definition 1.6.1, Λ_f^* is an outer measure on $\mathbb{R}$, which is said to be the L–S outer measure associated with f. The measure induced by Λ_f^* is denoted by Λ_f and said to be the L–S measure associated with f.

Theorem 1.6.4. *The outer measure Λ_f^* is a regular metric outer measure.*

Proof.

(1) Prove that Λ_f^* is a metric outer measure.

Let $E_1, E_2 \subset \mathbb{R}$ and $\rho(E_1, E_2) = \delta > 0$. For notational convenience, Λ_f^* is denoted by Λ^*. We will prove that

$$\Lambda^*(E_1 \cup E_2) \geq \Lambda^*(E_1) + \Lambda^*(E_2).$$

Obviously, we only need to consider the case $\Lambda^*(E_1 \cup E_2) < \infty$. For any $\varepsilon > 0$, there exists a sequence $I_k = (a_k, b_k]$ such that

$$\bigcup_{k=1}^{\infty} (a_k, b_k] \supset (E_1 \cup E_2),$$

$$\sum_{k=1}^{\infty} [f(b_k) - f(a_k)] < \Lambda^*(E_1 \cup E_2) + \varepsilon.$$

Divide each I_k into finite number of half-open congruent intervals $(c_j, d_j]$ satisfying $d_j - c_j < \delta$, $j = 1, \ldots, m = m_k$. Then

$$\bigcup_{j=1}^{\infty} (c_j, d_j] = I_k,$$

$$\sum_{j=1}^{\infty} [f(d_j) - f(c_j)] = f(b_k) - f(a_k).$$

Therefore, without loss of generality, the length $(b_k - a_k)$ of each interval I_k can be assumed to be less than σ. Let

$$N_1 = \{k \in \mathbb{N} : I_k \cap E_1 \neq \emptyset\} \quad \text{and} \quad N_2 = \{k \in \mathbb{N} : I_k \cap E_2 \neq \emptyset\},$$

then $N_1 \cap N_2 = \emptyset$ since $|I_k| < \rho(E_1, E_2), k \in \mathbb{N}$. Hence,

$$\bigcup_{k \in N_1} I_k \supset E_1, \quad \bigcup_{k \in N_2} I_k \supset E_2.$$

Thereby,

$$\Lambda^*(E_1 \cup E_2) + \varepsilon \geq \sum_{k \in N_1}^{\infty} [f(b_k) - f(a_k)] + \sum_{k \in N_2}^{\infty} [f(b_k) - f(a_k)]$$

$$\geq \Lambda^*(E_1) + \Lambda^*(E_2).$$

(2) Next, prove Λ^* is regular.

What we need to show is that for every $E \subset \mathbb{R}$, there exists a $B \in \mathscr{B}$ such that $E \subset B$ and $\Lambda^*(E) = \Lambda(B)$. For any $k \in \mathbb{N}$, take a sequence of half-open intervals $\{(a_i, b_i]\}$ such that $\cup_{i=1}^{\infty} (a_i, b_i] \supset E$ and

$$\sum_{i=1}^{\infty} [f(b_i) - f(a_i)] < \Lambda^*(E) + \frac{1}{k}.$$

Let $B_k = \cup_{i=1}^{\infty} (a_i, b_i]$ and $B = \cap_{k=1}^{\infty} B_k$. It is clear that $E \subset B \in \mathscr{B}$ and for any $k \in \mathbb{N}$, $\Lambda^*(E) \leq \Lambda^*(B) \leq \Lambda^*(B_k)$. Note that

$$\Lambda^*(B_k) \leq \sum_{i=1}^{\infty} [f(b_i) - f(a_i)] < \Lambda^*(E) + \frac{1}{k},$$

then $\Lambda^*(B) \leq \Lambda^*(E)$. As a result, $\Lambda^*(E) = \Lambda(B)$.   □

1.6.3   *The Lebesgue–Stieltjes Integral*

Let f be a real-valued monotonic increasing function and Λ^* be the L–S outer measure associated with f which induces the measure space $(\mathbb{R}, \mathscr{A}, \Lambda)$. It has been shown that the σ-algebra $\mathscr{A}$, consisting of all Λ^*-measurable sets, contains the σ-algebra $\mathscr{B}$ which is consisting of all Borel sets, and $(\mathbb{R}, \mathscr{B}, \Lambda)$ is also a measure space if Λ is restricted to $\mathscr{B}$. See from the proof of Theorem 1.6.4, for any $A \in \mathscr{A}$, there exists a $B \in \mathscr{B}$ satisfying

$$A = B \backslash Z, \ \Lambda(Z) = 0,$$

which implies that $\mathscr{A}$ and $\mathscr{B}$ are not much different, but the structure of $\mathscr{B}$ is much clearer than that of $\mathscr{A}$.

When we talk about the L–S integrals, it denotes the integrals on the measure space $(\mathbb{R}, \mathscr{A}, \Lambda)$ or $(\mathbb{R}, \mathscr{B}, \Lambda)$, where Λ is an L–S measure associated with a finite real-valued monotonic increasing function.

Theorem 1.6.5. *Let f be a real-valued monotonic increasing right continuous function on $\mathbb{R}$ and g be a bounded $\mathscr{B}$-measurable function on $\mathbb{R}$. If the Riemann–Stieltjes integral $\int_a^b g\,df$ on $[a,b]$ exists, then*

$$\int_{\mathbb{R}} g\chi_{(a,b)}\,d\Lambda_f = \int_a^b g\,df.$$

Proof.

(A) First, prove that for a finite interval $(c,d]$,

$$\Lambda_f((c,d]) = f(d) - f(c) = \lambda_f((c,d]).$$

In the following, the subscript f of Λ_f is dropped. Obviously,

$$\Lambda((c,d]) = \Lambda^*((c,d]) \le f(d) - f(c).$$

For any given $\varepsilon > 0$, there exists a sequence $I_k = (a_k, b_k]$ covering $(c,d]$ satisfying

$$\sum_{k=1}^{\infty} [f(b_k) - f(a_k)] < \Lambda((c,d]) + \varepsilon.$$

Due to the right continuity of f, there exists a $b_k' > b_k$ such that

$$f(b_k) \le f(b_k') < f(b_k) + 2^{-k}\varepsilon.$$

For any $a \in (c,d)$, it is clear that $\{I_k\}$ covers $[a,d]$. Thus, $\{(a_k, b_k')\}$ covers $[a,d]$. The finite covering theorem tells us there exists an $N \in \mathbb{N}$ such that $\cup_{k=1}^{N}(a_k, b_k') \supset [a,d]$, where the locally compact property of $\mathbb{R}$ is used. Hence,

$$\sum_{k=1}^{N} [f(b_k') - f(a_k)] \ge f(d) - f(a).$$

Note that

$$\sum_{k=1}^{N} [f(b_k') - f(a_k)] \le \sum_{k=1}^{\infty} [f(b_k) - f(a_k)] + \varepsilon,$$

therefore

$$f(d) - f(a) \leq \sum_{k=1}^{\infty} [f(b_k) - f(a_k)] + \varepsilon < \Lambda((c,d]) + 2\varepsilon.$$

Let $a \to c^+$ and $\varepsilon \to 0$,

$$f(d) - f(c) \leq \Lambda((c,d]).$$

(B) Next, let T be a partition of $[a, b]$:

$$x_0 = a < x_1 < \cdots < x_n = b.$$

The corresponding R–S upper and lower sums are

$$U(T) = \sum_{i=1}^{n} M_i[f(x_i) - f(x_{i-1})] \quad \text{and}$$

$$L(T) = \sum_{i=1}^{n} m_i[f(x_i) - f(x_{i-1})],$$

respectively, where

$$M_i = \sup\{g(x) : x \in [x_{i-1}, x_i]\}, \quad m_i = \inf\{g(x) : x \in [x_{i-1}, x_i]\}.$$

Let

$$g_1 = \sum_{i=1}^{n} M_i \chi_{(x_{i-1}, x_i]} \quad \text{and} \quad g_2 = \sum_{i=1}^{n} m_i \chi_{(x_{i-1}, x_i]},$$

then

$$g_1 \geq g \geq g_2, \int_{\mathbb{R}} g_1 d\Lambda \geq \int_{\mathbb{R}} g\chi_{[a,b]} d\Lambda \geq \int_{\mathbb{R}} g_2 d\Lambda.$$

According to (A), $\Lambda((x_{i-1}, x_i]) = f(x_i) - f(x_{i-1})$. Combined with the definition of the integral of simple function, it is known that

$$U(T) = \int_{\mathbb{R}} g_1 d\Lambda, \quad L(T) = \int_{\mathbb{R}} g_2 d\Lambda,$$

which implies that

$$U(T) \geq \int_{\mathbb{R}} g\chi_{[a,b]} d\Lambda \geq L(T).$$

Note that g is R–S integrable on $[a, b]$ with respect to f. Therefore, both $U(T)$ and $L(T)$ tend to $\int_a^b g df$ when the modulus of the partition T tends to 0. Thus,

$$\int_a^b gdf = \int_{\mathbb{R}} g\chi_{[a,b]}d\Lambda.$$

$\square$

Sections 1.6.2 and 1.6.3 give an example of a measure induced by an outer measure and the relationship between L–S integrals and R–S integrals, which has been discussed in theory of functions of real variable, under certain conditions. In this example, when f is the identity function, it coincides with the usual Lebesgue measure. The following section provides another example of a measure induced by an outer measure, which is just the Hausdorff outer measure on $\mathbb{R}^n$.

1.6.4 *Hausdorff Outer Measure*

Let $\alpha > 0, \varepsilon > 0$. For $A \subset \mathbb{R}^n$ with $A \neq \emptyset$, define

$$H_\alpha^\varepsilon(A) = \inf\left\{\sum_{k=1}^\infty \delta(A_k)^\alpha : \bigcup_{k=1}^\infty A_k \supset A, \delta(A_k) < \varepsilon\right\},$$

where $\delta(A_k)$ is the diameter of the point set A_k, that is,

$$\delta(A_k) = \sup\{|x - y| : x, y \in A_k\}.$$

Define

$$H_\alpha(A) = \lim_{\varepsilon \to 0^+} H_\alpha^\varepsilon(A), A \subset \mathbb{R}^n, A \neq \emptyset,$$

and

$$H_\alpha(\emptyset) = 0.$$

It is easy to verify that H_α is a regular metric outer measure on $\mathbb{R}^n$, which is called the Hausdorff outer measure. The measure induced by the Hausdorff outer measure is also named after the surname of Hausdorff. Note that H_α is equivalent to the Lebesgue outer measure m^* when $\alpha = n$, which means that there exist c and c' with $c > c' > 0$ satisfying $cH_n(A) \geq m^*(A) \geq c'H_n(A)$ for any $A \subset \mathbb{R}^n$. As for $\alpha > n$, $H_\alpha = 0$. We omit the details about H_α, and readers interested in H_α can refer to *Measure and Integrals* (Wheeden & Zygmund).

1.6.5   Induce an Outer Measure from a Measure and Then Induce the Extended Measure

The measures we talked about earlier are all defined over some σ-algebras. If the countable additivity of an σ-algebra is removed and only the finite additivity is guaranteed, it becomes an algebra. On an algebra, a premeasure is a generalized real-valued set function with the same properties as a measure. In fact, a premeasure is a measure. This section aims to describe the extension of a premeasure to a measure. More precisely, a premeasure can induce an outer measure, which in turn induces a measure that is an extension of the original premeasure.

Definition 1.6.6. A nonempty collection $\mathscr{A}$ of subsets of a set X is called an algebra on X provided

(1) $A \in \mathscr{A} \Rightarrow \complement A \in \mathscr{A}$;
(2) $A_1, \ldots, A_k \in \mathscr{A}, k \in \mathbb{N} \Rightarrow \cup_{i=1}^{k} A_i \in \mathscr{A}$.

Definition 1.6.7. Let $\mathscr{A}$ be a nonempty collection of subsets of a set X and μ be a nonnegative generalized real-valued function on $\mathscr{A}$. Then μ is called a premeasure on $(X, \mathscr{A})$ or $\mathscr{A}$ provided

(1) $\mu(\emptyset) = 0$;
(2) if $A_k \in \mathscr{A}$ for $k \in \mathbb{N}$, $A_k \cap A_j = \emptyset$ for $k \neq j$ and $\cup_{k=1}^{\infty} A_k \in \mathscr{A}$, then

$$\mu\left(\bigcup_{k=1}^{\infty} A_k \right) = \sum_{k=1}^{\infty} \mu(A_k).$$

The concepts of finite and σ-finite for the premeasure μ are the same as Definition 1.1.4.

Definition 1.6.8. Let λ be a premeasure on $(X, \mathscr{A})$. For any $E \subset X$,

$$\lambda^*(E) = \inf \left\{ \sum_{k=1}^{\infty} \lambda(A_k) : A_k \in \mathscr{A}, \bigcup_{k=1}^{\infty} A_k \supset E \right\},$$

is obviously an outer measure, which is said to be induced by the premeasure λ.

Theorem 1.6.6. *Let λ be a premeasure on $(X, \mathscr{A})$, λ^* be an outer measure induced by λ and $\mathscr{A}^*$ be a collection of λ^*-measurable sets. Then*

(1) $\lambda^*(A) = \lambda(A)$ *for all $A \in \mathscr{A}$;*
(2) $\mathscr{A} \subset \mathscr{A}^*$.

The proof of this theorem is easy and is left to the reader.

Definition 1.6.9. Let λ be a premeasure on $(X, \mathscr{A})$ and μ be a measure on a σ-algebra Σ which contains $\mathscr{A}$. The measure μ is called an extension of λ (from $\mathscr{A}$ to Σ) provided μ is the same as λ on $\mathscr{A}$.

Theorem 1.6.7 (Carathéodory–Hahn's extension theorem).
Let λ be a premeasure on $(X, \mathscr{A})$, λ^* be an outer measure induced by λ and $\mathscr{A}^*$ be the collection of λ^*-measurable sets, then

(1) λ^* *is an extension of λ (from $\mathscr{A}$ to $\mathscr{A}^*$);*
(2) *if λ is σ-finite, then, for any σ-algebra $\mathscr{M}$ satisfying $\mathscr{A} \subset \mathscr{M} \subset \mathscr{A}^*$, the extension from $\mathscr{A}$ to $\mathscr{M}$ is unique, which is the restriction of λ^* to $\mathscr{M}$.*

Proof. (1) is Theorem 1.6.6.
Let μ be the extension of λ from $\mathscr{A}$ to $\mathscr{M}$. Since λ is σ-finite, X can be rewritten as a disjoint union:

$$X = \bigcup_{k=1}^{\infty} S_k, \quad S_k \in \mathscr{A}, \quad \lambda(S_k) < \infty, \quad k \in \mathbb{N}.$$

For $E \in \mathscr{M}$, suppose that $E_k = E \cap S_k$, $k \in \mathbb{N}$, and for any sequence of $A_j \in \mathscr{A}$ with $E \subset \cup_{j=1}^{\infty} A_j$,

$$\mu(E_k) \leq \sum \mu(A_j) = \sum \lambda(A_j).$$

Thus,

$$\mu(E_k) \leq \lambda^*(E_k), \quad k \in \mathbb{N}.$$

Note that

$$\mu(S_k \backslash E_k) \leq \lambda^*(S_k \backslash E_k),$$
$$\mu(E_k) + \mu(S_k \backslash E_k) = \mu(S_k) = \lambda(S_k) = \lambda^*(S_k)$$
$$= \lambda^*(E_k) + \lambda^*(S_k \backslash E_k) < \infty.$$

Therefore,

$$\mu(E_k) = \lambda^*(E_k), \quad k \in \mathbb{N},$$

$$\mu(E) = \sum_{k=1}^{\infty} \mu(E_k) = \sum_{k=1}^{\infty} \lambda^*(E_k) = \lambda^*(E).$$

$\square$

Theorem 1.6.7, combined with the discussion in Section 1.6.2, yields the following two corollaries.

Corollary 1.6.8. *If measure spaces* $(\mathbb{R}, \mathcal{B}, \mu)$ *and* $(\mathbb{R}, \mathcal{B}, \nu)$ *satisfy*

$$\mu((a, b]) = \nu((a, b]) < \infty,$$

where $(a, b]$ *is any finite interval, then* $\mu = \nu$.

Corollary 1.6.9. *Let* $(\mathbb{R}, \mathcal{B}, \mu)$ *be a measure space with measure* μ *taking finite values on finite intervals, then there exists a monotonic increasing right continuous real-valued function* f *such that the L–S measure associated with* f *is the same as* μ *on* $\mathcal{B}$.

The detailed proofs of these two corollaries are left to readers. Here, we only provide the proof ideas. Let $\mathscr{A}$ be the smallest algebra containing all half-open intervals $(a, b]$, then $\mathscr{A}$ actually consists of the empty set and finite union of intervals in the form of $(-\infty, a], (a, b]$ and $(b, +\infty)$. Corresponding to the measure μ on $\mathcal{B}$, define

$$f(x) = \begin{cases} \mu((0, x]) & \text{if } x > 0, \\ 0 & \text{if } x = 0, \\ -\mu((0, x]) & \text{if } x < 0. \end{cases}$$

It is easy to verify that f is a monotonic increasing right continuous real-valued function. Define λ_f as in Section 1.6.2 and allow a and b to take infinite values in the equation $\lambda_f((a, b]) = f(b) - f(a)$, then it can be verified that this λ_f is a premeasure on $\mathscr{A}$. Therefore, by Theorem 1.6.7, the two corollaries can be obtained.

Exercise 1.6

1. Let μ^* be a regular outer measure on X, $E_n \subset X$ and $E_n \subset E_{n+1}$. Show that

$$\lim_{n\to\infty} \mu^*(E_n) = \mu^* \left(\lim_{n\to\infty} E_n \right).$$

2. Prove Theorem 1.6.1.

3. Show that Theorem 1.6.5 fails when f is not right continuous. Please give a counterexample.

4. Prove that H_α defined in Section 1.6.4 is indeed a metric outer measure on $\mathbb{R}^n$ and is regular.

5. Prove Theorem 1.6.6 which is about measure extension.

6. Suppose f is a monotonic increasing and absolutely continuous function, Λ_f is an L–S measure associated with f, then $\Lambda_f \ll m$ where m denotes the usual Lebesgue measure.

7. Let I be the identity function on $\mathbb{R}$, that is, $I(x) = x$. Then $\Lambda_I = m$ where m is the usual Lebesgue measure.

8. Suppose $(X, \mathscr{A}, \mu)$ is a measure space induced by an outer measure μ^* on X and μ_1^* is the outer measure induced by μ. Show that

   (1) $\mu_1^* \geq \mu^*$;
   (2) for a set $E \subset X$, $\mu^*(E) = \mu_1^*(E)$ if and only if there exists an $A \in \mathscr{A}$ such that $E \subset A$ and $\mu(A) = \mu^*(E)$.

   Hence, $\mu_1^* = \mu^*$ provided μ^* is regular.

9. Construct an irregular outer measure.

10. Let Λ_f be an L–S measure induced by a monotonic increasing function f on $\mathbb{R}$. Show that if $\Lambda_f \ll m$ where m denotes the usual Lebesgue measure, then

$$\frac{d\Lambda_f}{dm} = f' \ (m\text{-a.e.}).$$

11. Prove Corollaries 1.6.8 and 1.6.9.

1.7 Product Measure and Fubini's Theorem

In theory of functions of real variable, we have learned the method of calculating multiple integrals using repeated integrals which is based on the famous Fubini's theorem. Fubini's theorem is one of the most important results in analysis, which not only has significant theoretical significance but also has high practical value. For example, in the theory of Fourier transform, Fubini's theorem is of fundamental importance. This theorem is also widely used in many occasions in probability theory.

1.7.1 Some Simple Examples

Example 1.7.1. Let $I = (0, 1), I^2 = (0, 1) \times (0, 1)$ and

$$f(x, y) = \frac{x^2 - y^2}{(x^2 + y^2)^2} \quad (x, y) \in I^2.$$

As a Riemann integral,

$$\int_I f(x, y) dy = \int_0^1 \left(\frac{y}{x^2 + y^2} \right)'_y dy = \frac{1}{1 + x^2}, \quad x \in I.$$

Then, as a Lebesgue integral,

$$\int_I \left(\int_I f(x, y) dy \right) dx = \frac{\pi}{4}$$

and

$$\int_I \left(\int_I f(x, y) dx \right) dy = -\frac{\pi}{4}.$$

We find that the result of integrating f with respect to x first and then with respect to y is different from that of integrating with respect to y first and then with respect to x. We point out an important fact: $f \notin L(I^2)$. To prove $f \notin L(I^2)$, for any $k \in \mathbb{N}$, define

$$D_k = \{(x, y) \in I^2 : 2^{-k} < x^2 + y^2 < 2^{-k+1}, \ x > 2y\}.$$

It is obvious that $D_k, k \in \mathbb{N}$ are pairwise disjoint. On D_k,

$$f(x, y) = \left(1 - \frac{2y^2}{x^2 + y^2} \right) \frac{1}{x^2 + y^2}$$

$$> \left(1 - \frac{2y^2}{4y^2 + y^2} \right) \frac{1}{x^2 + y^2} > \frac{3}{5} \cdot 2^{k-1}.$$

Note that the Lebesgue measure of D_k is

$$|D_k| = \pi(2^{-k+1} - 2^{-k})\theta = \pi\theta 2^{-k},$$

where

$$\theta = \frac{1}{2\pi}\arctan\frac{1}{2}.$$

Therefore,

$$\int_{I^2} |f|dxdy \geq \sum_{k=1}^{\infty}\int_{D_k}|f|dxdy \geq \sum_{k=1}^{\infty}\left(\frac{3}{5}\cdot 2^{k-1}\cdot\pi\theta 2^{-k}\right) = \infty.$$

Example 1.7.2. Let $f(x,y) = \frac{xy}{(x^2+y^2)^2}, (x,y) \neq (0,0)$. If $x \neq 0$, then $f(x,\cdot) \in L(\mathbb{R})$ and

$$\int_{\mathbb{R}} f(x,y)dy = 0.$$

If $y \neq 0$, then $f(\cdot,y) \in L(\mathbb{R})$ and

$$\int_{\mathbb{R}} f(x,y)dx = 0.$$

Therefore,

$$\int_{\mathbb{R}}\left(\int_{\mathbb{R}} f(x,y)dy\right)dx = \int_{\mathbb{R}}\left(\int_{\mathbb{R}} f(x,y)dx\right)dy = 0.$$

But $f \notin L(\mathbb{R}^2)$. In order to prove this, for every $k \in \mathbb{N}$, define

$$S_k = \left\{(x,y) \in \mathbb{R}^2 : 2^{-k} < x^2 + y^2 < 2^{-k+1}, 0 < \frac{1}{2}y < x < 2y\right\}.$$

It is obvious that $S_k, k \in \mathbb{N}$ are pairwise disjoint. On S_k, $xy > \frac{1}{4}(x^2 + y^2)$, thus

$$f(x,y) > \frac{1}{4}\frac{1}{x^2 + y^2} > 2^{k-3}.$$

Note that the Lebesgue measure of S_k is

$$|S_k| = \pi\theta 2^{-k},$$

where

$$\theta = \frac{1}{2\pi} \arctan \frac{3}{4}.$$

Therefore,

$$\int_{\mathbb{R}^2} |f| dx dy \geq \sum_{k=1}^{\infty} \int_{S_k} f dx dy \geq \sum_{k=1}^{\infty} 2^{k-3} \cdot \pi \theta 2^{-k} = \infty.$$

Example 1.7.3. Suppose $I = [0,1] \times [0,1]$, μ_1 is the usual Lebesgue measure, μ_2 is the counting measure on $[0,1]$ and $\Delta = \{(x,y) : 0 \leq x \leq 1\}$. Let $f(x,y) = \chi_\Delta(x,y)$, then

$$\int_{[0,1]} f(x,y) d\mu_1(x) = 0 \quad \text{for all } y \in [0,1],$$

$$\int_{[0,1]} f(x,y) d\mu_2(y) = 1 \quad \text{for all } x \in [0,1],$$

thus

$$\int_{[0,1]} \left(\int_{[0,1]} f(x,y) d\mu_1(x) \right) d\mu_2(y) = 0,$$

$$\int_{[0,1]} \left(\int_{[0,1]} f(x,y) d\mu_2(y) \right) d\mu_1(x) = 1.$$

From the above examples, it can be seen that it is quite complex to exchange the order of repeated integrals. An important way to solve this problem is to introduce the concept of product measure.

1.7.2 Product Measure

Definition 1.7.1. Suppose $(X, \mathscr{A}, \mu)$ and $(Y, \mathscr{B}, \nu)$ are two measure spaces. Let $X \times Y$ denote the product of X and Y, that is,

$$X \times Y = \{(x,y) : x \in X, y \in Y\}.$$

(1) If $A \subset X$ and $B \subset Y$, then $A \times B$ is called a rectangle in $X \times Y$. If $A \in \mathscr{A}$ and $B \in \mathscr{B}$, then $A \times B$ is called a measurable rectangle. Let $\mathscr{R}$ represent the collection of all measurable rectangles and $\mathscr{A} \times \mathscr{B}$ denote the smallest σ-algebra containing $\mathscr{R}$ on $X \times Y$.

(2) Define an area function λ:

$$\lambda(A \times B) = \mu(A)\nu(B), \quad A \times B \in \mathscr{R}.$$

Recall that the product of zero and infinity is equal to zero. The product measure is induced by an outer measure on $X \times Y$ based on the area function λ of the measurable rectangle. This can be done from the perspective of measure extension by using results in Section 1.6.5. However, we prefer to omit the overly formal discussion in Section 1.6.5 and define the outer measure directly in terms of λ because it is much more concise and acceptable. Of course, the discussion in Section 1.6.5 has its own theoretical significance. Interested readers can read it for themselves. In the following discussion, it will be assumed that $(X, \mathscr{A}, \mu)$ and $(Y, \mathscr{B}, \nu)$ are measure spaces and $\mathscr{R}$ is the collection of all measurable rectangles in $X \times Y$.

Definition 1.7.2. For any $E \subset X \times Y$, define

$$\lambda^*(E) = \inf \left\{ \sum_{k=1}^{\infty} \lambda(I_k) : \ I_k \in \mathscr{R}, \ \bigcup_{k=1}^{\infty} I_k \supset E \right\}.$$

Lemma 1.7.1. *The measure λ^* is an outer measure on $X \times Y$ and*

$$\lambda^*(I) = \lambda(I) \quad \text{for all } I \in \mathscr{R}.$$

Proof. According to Definition 1.6.1, it is only necessary to prove that λ^* meets the countable subadditivity. Let $E_k \subset X \times Y$, $k \in \mathbb{N}$ and $E = \cup_{k=1}^{\infty} E_k$. On the base of the definition of λ^*, for any $\varepsilon > 0$, there exist measurable rectangles I_{kj}, $k \in \mathbb{N}$, $j \in \mathbb{R}$ such that

$$\bigcup_{j=1}^{\infty} I_{kj} \supset E_k, \quad \sum_{j=1}^{\infty} \lambda(I_{kj}) \leq \lambda^*(E_k) + 2^{-k}\varepsilon.$$

Obviously,

$$\bigcup_{k=1}^{\infty} \bigcup_{j=1}^{\infty} I_{kj} \supset E.$$

Hence,

$$\lambda^*(E) \leq \bigcup_{k=1}^{\infty} \bigcup_{j=1}^{\infty} \lambda(I_{kj}) \leq \sum_{k=1}^{\infty} \lambda^*(E_k) + \varepsilon,$$

which shows the countable subadditivity of λ^*. $\qquad\square$

The collection of all λ^*-measurable sets, that is, the sets with respect to λ^* satisfying the Carathéodory condition which can be seen in Definition 1.6.2, is denoted by $\mathscr{M}$ and the restriction of λ^* to $\mathscr{M}$ is denoted by λ, that is, $\lambda = \mu \times \nu$. By Theorem 1.6.1, $(X \times Y, \mu, \lambda)$ is a complete measure space.

Lemma 1.7.2.

$$\mathscr{A} \times \mathscr{B} \subset \mathscr{M}.$$

Proof. It is only necessary to prove that each measurable rectangle $I = A \times B$ satisfies the Carathéodory condition with respect to the outer measure λ^*. Let

$$I_1 = (X \backslash A) \times B, \quad I_2 = A \times (Y \backslash B), \quad I_3 = (X \backslash A) \times (Y \backslash B).$$

Obviously, I_1, I_2 and I_3 are all measurable rectangles and $\complement I = I_1 \cup I_2 \cup I_3$. For any $E \subset X \times Y$ and $\varepsilon > 0$, there exist measurable rectangles J_k, $k \in \mathbb{N}$ such that

$$\bigcup_{k=1}^{\infty} J_k \supset E, \quad \bigcup_{k=1}^{\infty} \lambda(J_k) \leq \lambda^*(E) + \varepsilon.$$

Let $J_{k0} = J_k \cap I, \quad J_{k\nu} = J_k \cap I_\nu, \quad \nu = 1, 2, 3$, then

$$\bigcup_{k=1}^{\infty} J_{k0} \supset E \cap I \quad \bigcup_{\nu=1}^{3} \bigcup_{k=1}^{\infty} J_{k\nu} \supset E \cap \complement I.$$

Therefore,

$$\lambda^*(E \cap I) \leq \sum_{k=1}^{\infty} \lambda(J_{k0}), \quad \lambda^*(E \cap \complement I) \leq \sum_{\nu=1}^{3} \sum_{k=1}^{\infty} \lambda(J_{k\nu}).$$

Hence,

$$\lambda^*(E \cap I) + \lambda^*(E \cap \complement I) \leq \sum_{k=1}^{\infty} \sum_{\nu=1}^{3} \lambda(J_{k\nu}) = \sum_{k=1}^{\infty} \lambda(J_k) \leq \lambda^*(E) + \varepsilon,$$

which indicates that I satisfies the Carathéodory condition. $\square$

Definition 1.7.3. The above $(X \times Y, \mathscr{M}, \lambda)$ and $(X \times Y, \mathscr{A} \times \mathscr{B}, \lambda)$ are both called the product measure spaces of $(X, \mathscr{A}, \mu)$ and $(Y, \mathscr{B}, \nu)$. Correspondingly, $\lambda = \mu \times \nu$ is called the product measure.

Remark. The measure space $(X \times Y, \mathscr{M}, \mu \times \nu)$ is induced by the outer measure and consequently must be complete. However, $(X \times Y, \mathscr{A} \times \mathscr{B}, \mu \times \nu)$ can be incomplete. Indeed, if there exists an $A \subset X$ such that $A \notin \mathscr{A}$ and a $B \in \mathscr{B}$ with $B \neq \emptyset$ and $\nu(B) = 0$, then $A \times B \notin \mathscr{A} \times \mathscr{B}$. But $X \times B \in \mathscr{A} \times \mathscr{B}$ and

$$\mu \times \nu(A \times B) = \mu \times \nu(X \times B) = 0.$$

1.7.3 Fubini's Theorem

If $E \subset X \times Y$, for $x \in X$ and $y \in Y$, define the x-section E_x and the y-section E^y of E by

$$E_x = \{y \in Y : (x, y) \in E\}, \quad E^y = \{x \in X : (x, y) \in E\}.$$

Lemma 1.7.3. *If $E \in \mathscr{A} \times \mathscr{B}$, then $E_x \in \mathscr{B}$ for $x \in X$ and $E^y \in \mathscr{A}$ for $y \in Y$.*

Proof. Define

$$\mathscr{F} = \{E \in \mathscr{A} \times \mathscr{B} : E_x \in \mathscr{B} \text{ for all } x \in X\},$$

then $\mathscr{F} \supset \mathscr{R}$. It is apparent that $\emptyset \in \mathscr{F}$ and $X \times Y \in \mathscr{F}$. Let $E_n \in \mathscr{F}, n \in \mathbb{N}$, then

$$\left(\bigcup_{n=1}^{\infty} E_n \right)_x = \bigcup_{n=1}^{\infty} (E_n)_x \in \mathscr{B},$$

which shows that $\mathscr{F}$ is closed under countable unions. If $E \in \mathscr{F}$, then

$$(\complement E)_x = \complement(E_x) \in \mathscr{B},$$

which indicates that $\mathscr{F}$ is closed under complements. Therefore, $\mathscr{F}$ is a σ-algebra and $\mathscr{F} \supset \mathscr{A} \times \mathscr{B}$ since $\mathscr{A} \times \mathscr{B}$ is the smallest σ-algebra containing $\mathscr{R}$. $\qquad \square$

Lemma 1.7.4. *Let f be a generalized real-valued $\mathscr{A} \times \mathscr{B}$-measurable function defined on $X \times Y$, then*

(1) *as a function on* Y, $f(x, \cdot)$ *is* $\mathscr{B}$-*measurable for any* $x \in X$;
(2) *as a function on* X, $f(\cdot, y)$ *is* $\mathscr{A}$-*measurable for any* $y \in Y$.

Proof. Verify only (1). Let f be the characteristic function of $E \in \mathscr{A} \times \mathscr{B}$, then $\chi_E(x, y) = \chi_{E_x}(y)$ for any $x \in X$. According to Lemma 1.7.3, $E_x \in \mathscr{B}$, thus $\chi_{E_x}(y)$ is $\mathscr{B}$-measurable, which implies that (1) is true for simple functions. Hence, (1) holds for $\mathscr{A} \times \mathscr{B}$-measurable functions since a measurable function must be the limit of a sequence of simple functions. $\qquad\square$

Lemma 1.7.5. *Let* μ *and* ν *be complete and* $\mathscr{R}_{\sigma\delta}$ *be the collection of all sets in the form of* $\cap_{i=1}^{\infty} \cap_{j=1}^{\infty} E_{i,j}(E_{i,j} \in \mathscr{R})$. *If* $E \in \mathscr{R}_{\sigma\delta}$ *and* $\mu \times \nu(E) < \infty$, *then*

(1) *the function* $\Gamma_E(x) := \nu(E_x)$ *is* $\mathscr{A}$-*measurable;*
(2) *a the function* $\Gamma^E(y) := \mu(E^y)$ *is* $\mathscr{B}$-*measurable;*
(3) $(\mu \times \nu)(E) = \int_Y \mu(E^y) d\nu(y) = \int_X \nu(E_x) d\mu(x)$.

Proof. If $E \in \mathscr{R}$, then the conclusions are right. Let $\mathscr{R}_\sigma$ be the collection of all sets which can be represented as countable unions of elements in $\mathscr{R}$. Suppose $E \in \mathscr{R}_\sigma$, $E = \cup_{j=1}^{\infty} E_j$, $E_j \in \mathscr{R}$, it can be considered that $E_i \cap E_k = \emptyset$ for $i \neq k$. Since $\nu(E_x) = \sum_{j=1}^{\infty} \nu((E_j)_x)$,

$$\int_X \nu(E_x) d\mu = \sum_{j=1}^{\infty} \int_X \nu((E_j)_x) d\mu = \sum_{j=1}^{\infty} (\mu \times \nu)(E_j) = (\mu \times \nu)(E).$$

Therefore, the conclusions hold for $E \in \mathscr{R}_\sigma$.

Let $E \in \mathscr{R}_{\sigma\delta}$ and $(\mu \times \nu)(E) < \infty$, then $E = \cap_{j=1}^{\infty} E_j$ for $E_j \in \mathscr{R}_\sigma$. It can be considered that $E_j \supset E_{j+1}$ and $(\mu \times \nu)(E_1) < \infty$. Due to the fact that $E_1 \in \mathscr{R}_\sigma$,

$$(\mu \times \nu)(E_1) = \int_X \nu((E_1)_x) d\mu < \infty.$$

Thus, $\nu((E_1)_x) < \infty$, μ-a.e. On account of $E_x = \cap_{j=1}^{\infty} (E_j)_x$,

$$\nu(E_x) = \lim_{j \to \infty} \nu((E_j)_x), \quad \mu\text{-a.e.}$$

holds at least at the point where $\nu((E_1)_x) < \infty$. Since μ is a complete measure,

$$\Gamma_E(x) := \nu(E_x)$$

is $\mathscr{A}$-measurable. The Lebesgue dominated convergence theorem shows that

$$\int_X \nu(E_x)d\mu = \lim_{j\to\infty} \int_X \nu((E_j)_x)d\mu = \lim_{j\to\infty} (\mu \times \nu)(E_j) = \mu \times \nu(E). \qquad \square$$

Remark. The condition $(\mu \times \nu)(E) < \infty$ in Lemma 1.7.5 is necessary. There exists a counterexample as follows. Let $X = Y = [0,1]$, $\mathscr{A} = \mathscr{B} = \{$the usual Lebesgue measurable sets of $[0,1]\}$. Define $\mu = \nu$ as follows:

$$\mu(\{x\}) = \begin{cases} 2 & \text{if } x \in A, \\ 1 & \text{if } x \notin A, \end{cases}$$

where A is a nonmeasurable subset of $[0,1]$ and $\mu = 0$ on the empty set. Obviously, μ is complete. Let

$$D = \{(x,y) \in [0,1]^2 : x = y\}$$

and

$$I_{n,k} = \left[\frac{k-1}{n}, \frac{k}{n}\right], \quad k = 1, 2, \ldots, n, \ n \in \mathbb{N},$$

then

$$D = \bigcap_{n=1}^{\infty} \bigcup_{k=1}^{n} (I_{n,k} \times I_{n,k}) \in \mathscr{R}_{\sigma\delta}.$$

It is clear that $\mu \times \nu(D) = \infty$. However,

$$\nu(D_x) = \chi_A(x) + 1 = \Gamma_D(x)$$

is nonmeasurable.

Lemma 1.7.6. *Let μ and ν be complete, $E \in \mathscr{M}$ and $(\mu \times \nu)(E) < \infty$, then there exists an $A_0 \in \mathscr{A}$ with $\mu(A_0) = 0$ such that $E_x \in \mathscr{B}$ provided $x \in X \backslash A_0$, the function*

$$\Gamma_E(x) = \begin{cases} 0 & \text{when } x \in A_0, \\ \nu(E_x) & \text{when } x \in X \backslash A_0 \end{cases}$$

is $\mathscr{A}$-measurable and

$$(\mu \times \nu)(E) = \int_X \Gamma_E(x)d\mu(x).$$

Proof.   According to the definition of $\mu \times \nu$, there exists an $F \in$ $\mathscr{R}_{\sigma\delta}$ with $F \supset E$ and $(\mu \times \nu)(F) = (\mu \times \nu)(E)$. Let $G = F\backslash E$, then $(\mu \times \nu)(G) = 0$. For G, there exists an $H \in \mathscr{R}_{\sigma\delta}$ such that $G \subset H$ and $(\mu \times \nu)(H) = 0$. In the light of Lemma 1.7.5, $\nu(H_x)$ is $\mathscr{A}$-measurable and

$$(\mu \times \nu)(H) = \int_X \nu(H_x)d\mu(x).$$

Therefore, $\nu(H_x) = 0, \mu$-a.e. Suppose $A_0 = \{x \in X : \nu(H_x) \neq 0\}$, then $\mu(A_0) = 0$. The set G_x is the subset of the ν-null set for $x \in$ $X\backslash A_0$ owing to the fact $G_x \subset H_x$. Further, G_x is the ν-null set for $x \in X\backslash A_0$ due to the completeness of ν. By Lemma 1.7.5, $F_x \in \mathscr{B}$, thus for any $x \in X\backslash A_0$,

$$E_x = F_x\backslash G_x \in \mathscr{B}$$

and Γ_E is $\mathscr{A}$-measurable. Applying Lemma 1.7.5 again, we obtain

$$\int_X \Gamma_E(x)d\mu(x) = \int_{X\backslash A_0} \nu(E_x)d\mu = \int_X \nu(F_x)d\mu = (\mu \times \nu)(E).$$

$\square$

Theorem 1.7.7 (Fubini's theorem).   *Let μ and ν be complete and* $f \in L(X \times Y, \mathscr{M}, \mu \times \nu)$, *then*

(1) *the x-section of f, $f(x, \cdot)$, is $\mathscr{B}$-measurable for almost all $x \in X$;*
(2) $\int_Y f(x,y)d\nu(y) \in L(X, \mathscr{A}, \mu)$;
(3) $\int_{X \times Y} f d(\mu \times \nu) = \int_X \left(\int_Y f(x,y)d\nu(y) \right)d\mu(x)$.

Proof.   In accordance with Lemma 1.7.6, if $E \in \mathscr{M}, (\mu \times \nu)(E) <$ ∞, then

$$\int_Y \chi_E(x,y)d\nu(y) = \nu(E_x), \quad \mu\text{-a.e.}$$

and

$$\int_X \nu(E_x)d\mu(x) = \mu \times \nu(E) = \int_{X \times Y} \chi_E d(\mu \times \nu),$$

which imply the conclusions of the theorem hold for the characteristic function χ_E and thus for nonnegative integrable simple functions.

Further, for nonnegative integrable functions, the limit of monotonic increasing nonnegative integrable simple functions, the conclusions are also true. Thereby, according to the Levi theorem, the three conclusions hold for all integrable functions. □

Theorem 1.7.8 (Tonelli's theorem). *Let μ and ν be complete and σ-finite and f be a nonnegative $\mathscr{M}$-measurable function, then*

(1) *the x-section of f, $f(x,\cdot)$, is $\mathscr{B}$-measurable for almost all $x \in X$;*
(2) *$\int_Y f(x,y)d\nu(y)$ is $\mathscr{A}$-measurable;*
(3) *$\int_{X \times Y} f d(\mu \times \nu) = \int_X \left(\int_Y f(x,y)d\nu(y) \right) d\mu(x)$.*

Proof. Since μ and ν are σ-finite, $\mu \times \nu$ is also σ-finite. Thus, the nonnegative $\mathscr{M}$-measurable function f can be expressed as the limit of monotonic increasing integrable simple functions. Therefore, similar to Theorem 1.7.7, this theorem also holds. □

Lemma 1.7.9. *If μ and ν are σ-finite, then the three conclusions of Lemma 1.7.5 hold.*

Proof. Let $\mathscr{F}$ be the collection of all $\mu \times \nu$-measurable sets, i.e. $\mathscr{M}$-measurable sets, that make the three conclusions of Lemma 1.7.5 hold. As proved in Lemma 1.7.5, $\mathscr{R}_\sigma \subset \mathscr{F}$.

Suppose μ and ν are both finite, then it is clear that $\mathscr{F}$ is closed under countable increasing unions and countable decreasing intersections. A nonempty collection of sets is called a monotone class if it is closed under countable increasing unions and countable decreasing intersections. Therefore, $\mathscr{F}$ is a monotone class.

Bearing the fact $\mathscr{R}_\sigma \subset \mathscr{F}$ in mind, the monotone class theorem, which is placed after the proof of Lemma 1.7.9, make known that $\mathscr{F}$ contains the σ-algebra generated by $\mathscr{R}$, that is, $\mathscr{A} \times \mathscr{B}$. Let μ and ν be σ-finite, then X and Y can be represented as pairwise disjoint unions, respectively. That is, $X = \cup_{i=1}^\infty X_i$, where $X_i, i \in \mathbb{N}^+$ are pairwise disjoint satisfying $\mu(X_i) \in \mathbb{R}$ and $Y = \cup_{j=1}^\infty Y_j$, where $Y_j, j \in \mathbb{N}^+$ are pairwise disjoint satisfying $\mu(Y_j) \in \mathbb{R}$. As a result,

$$X \times Y = \bigcup_{i=1}^\infty \bigcup_{j=1}^\infty X_i \times Y_j,$$

where $X_i \times Y_j$ are pairwise disjoint. For any $E \in \mathscr{A} \times \mathscr{B}$, define $E_{i,j} = E \cap (X_i \times Y_j)$. Apply the proved results to $E_{i,j}$, and then combine them to obtain the desired conclusions.   □

Appendix. The Monotone Class Theorem and its Proof

In Definition 1.1.1, the σ-ring has been defined. A ring is a nonempty collection of sets that is closed under differences and finite unions. A σ-ring is a ring closed under countable unions.

Theorem 1.7.10 (Monotone class theorem). *Let* $\sigma_r(\mathscr{R})$ *be a* σ-*ring generated by a ring* $\mathscr{R}$ *and* $\mathscr{M}(\mathscr{R})$ *be the monotone class generated by the ring* $\mathscr{R}$, *then*

$$\sigma_r(\mathscr{R}) = \mathscr{M}(\mathscr{R}).$$

Proof.   Note that $\sigma_r(\mathscr{R})$ is also a monotone class, therefore

$$\mathscr{M}(\mathscr{R}) \subset \sigma_r(\mathscr{R}).$$

For $A \in \mathscr{M}(\mathscr{R})$, define

$$\mathscr{M}_A = \{B \in \mathscr{M}(\mathscr{R}) : B \cup A, B \backslash A, A \backslash B \in \mathscr{M}(\mathscr{R})\}.$$

Clearly, $\mathscr{M}_A$ is a monotone class. It is easy to show $B \in \mathscr{M}_A \Leftrightarrow A \in \mathscr{M}_B$.

If $A \in \mathscr{R}$, then $\mathscr{R} \subset \mathscr{M}_A$ since $\mathscr{R}$ is a ring, thus $\mathscr{M}(\mathscr{R}) \subset \mathscr{M}_A$. Hence, $B \in \mathscr{M}_A$ for every $B \in \mathscr{M}(\mathscr{R})$, that is, $A \in \mathscr{M}_B$. Note that A is any element of $\mathscr{R}$, so $\mathscr{R} \subset \mathscr{M}_B$. Thereby, $\mathscr{M}(\mathscr{R}) \subset \mathscr{M}_B$ which means that $\mathscr{M}(\mathscr{R})$ is a ring, and thus, a σ-ring. Therefore,

$$\sigma_r(\mathscr{R}) \subset \mathscr{M}(\mathscr{R}).$$ □

Theorem 1.7.11 (Fubini–Tonelli's theorem). *Let* $(X, \mathscr{A}, \mu)$ *and* $(Y, \mathscr{B}, \nu)$ *be two* σ-*finite measure spaces and* f *be a generalized real-valued* $\mathscr{A} \times \mathscr{B}$-*measurable function on* $X \times Y$. *If* f *is integrable or nonnegative, then*

$$\int_{X \times Y} f d(\mu \times \nu) = \int_Y \left(\int_X f d\mu \right) d\nu = \int_X \left(\int_Y f d\nu \right) d\mu.$$

Proof.   Lemma 1.7.9 indicates that Theorem 1.7.11 holds for characteristic functions and thus for simple functions. Then the proof is completed by transitioning it to the limit.   □

Exercise 1.7

1. Let $X = Y = [0, 1]$, $\mathscr{A} = \mathscr{B}$ is a σ-algebra containing all open sets of X. Prove $\mathscr{A} \times \mathscr{B}$ contains all open sets of $X \times Y$.

2. Let $f \in L(X, \mathscr{A}, \mu), g \in L(Y, \mathscr{B}, \nu)$. Show that $f(x)g(y) \in L(X \times Y, \mathscr{A} \times \mathscr{B}, \mu \times \nu)$ and

$$\int f(x)g(y)d(\mu \times \nu) = \int f d\mu \cdot \int g d\nu.$$

3. For $x > 0$, $\int_{(0,\infty)} e^{-xt}dt = \frac{1}{x}$. Combining the results with Fubini's theorem to prove

$$\lim_{n \to \infty} \int_0^n \frac{\sin x}{x} dx = \frac{\pi}{2}.$$

4. Let f be a nonnegative real-valued function on $\mathbb{R}$. Based on the usual Lebesgue measure, prove that f is measurable if and only if $E = \{(x, y) : 0 \leq y \leq f(x)\}$ is a σ-algebra generated by the collection of all measurable rectangles of $\mathbb{R}^2$.

5. On the base of the conditions of Exercise 4, set f is integrable and μ is the usual Lebesgue measure. Show that

$$\mu \times \mu(E) = \int_{\mathbb{R}} f d\mu.$$

6. Consider the usual Lebesgue measure. Let $1 \leq p < \infty$ and f be measurable on $[0, 1] \times [0, 1]$. Prove that

$$\left\{ \int_{(0,1)} \left(\int_{(0,1)} |f(x, y)| dy \right)^p dx \right\}^{\frac{1}{p}}$$

$$\leq \int_{(0,1)} \left(\int_{(0,1)} |f(x, y)|^p dx \right)^{\frac{1}{p}} dy.$$

7. Let E be a Lebesgue measurable set on $\mathbb{R}$ and μ be a Lebesgue measure. Show that $\{(x, y) : x - y \in E\}$ is $\mu \times \mu$-measurable. Use the result to prove that if f is measurable on $\mathbb{R}$, then $f(x - y)$ is measurable on $\mathbb{R}^2$.

   Hint: Consider the case that E is an open set first, then a G_δ set and a null set.

8. Let f and g be Lebesgue integrable functions on $\mathbb{R}$, which can be complex-valued. Show that

$$(f * g)(x) := \frac{1}{2\pi} \int_{\mathbb{R}} f(x - y)g(y)dy$$

   is Lebesgue measurable on $\mathbb{R}$ by the Tonelli theorem, where $f * g$ is called the convolution of f and g.

9. Let f be a Lebesgue integrable function on $\mathscr{R}$ and define the Fourier transform of f by

$$\hat{f}(t) = \frac{1}{2\pi} \int_{\mathbb{R}} e^{-itx} f(x)dx,$$

   which is a complex-valued function. Prove that

   (1) $\hat{f} \in C(\mathbb{R})$ and $\lim_{t \to \infty} \hat{f}(t) = 0$;
   (2) $\widehat{f * g} = \hat{f} \cdot \hat{g}$.

Chapter 2

Measure and Topology

In this chapter, we study the measure and integral on locally compact Hausdorff topological space. Obviously, some contents related to topology in the theory of functions of real variable can only be generalized on topological space. For example, a measurable set can be approximated by open sets or compact sets. We will consider whether there is a similar conclusion for topological space. In addition, the continuous function plays an important role on the topological space. Therefore, we will study the continuous function on the locally compact Hausdorff space in Section 2.2. In Section 2.3, we study the relationship between the linear functional on $C_c(X)$ space and the measure on the topological space, which is an important link between the functional analysis and the measure theory. In Section 2.4, we will extend the Luzin theorem about the construction of measurable functions on $\mathbb{R}^n$ to the locally compact Hausdorff topological space. In Section 2.5, we briefly introduce the Radon product of measures. The concept of Haar measure is important when studying the analysis problems on topological groups. In Section 2.6, the definition of Haar measure will be given.

2.1 Topological Space and Continuous Mapping

In this section, we introduce some necessary knowledge about topological space used in this chapter, see [10, 11] for details. We mainly give the definition of continuous mapping and prove the Uryson extension theorem about continuous function.

Definition 2.1.1. Let X be a nonempty set. A topology on X is a family $\mathscr{T}$ of subsets of X satisfying three conditions:

(i) $\emptyset \in \mathscr{T}$, $X \in \mathscr{T}$;
(ii) if $\{U_\alpha\} \subset \mathscr{T}$, then $\bigcup_\alpha U_\alpha \in \mathscr{T}$;
(iii) if $\{U_i : i = 1, \ldots, m\} \subset \mathscr{T}$, then $\bigcap_{i=1}^m U_i \in \mathscr{T}$.

Definition 2.1.2. If $\mathscr{T}$ is a topology on X, the pair $(X, \mathscr{T})$ is called a topological space. Within context and in the absence of ambiguity, $(X, \mathscr{T})$ is simply written as the topological space X. The members of $\mathscr{T}$ are called open sets of X, and their complements are called closed sets.

Let $A \subset X$, $\overset{\circ}{A} = \bigcup_{V \subset A, V \in \mathscr{T}} V$ is called the interior of A. In other words, $\overset{\circ}{A}$ is the union of all open sets contained in A.

Let $A \subset X$, $\overline{A} = \bigcap_{F \supset A, \complement F \in \mathscr{T}} F$ is called the closure of A. In other words, $\overline{A}$ is the intersection of all closed sets containing A.

If there exists a family of sets $\{V_\alpha\}$ such that $\bigcup_\alpha V_\alpha \supset A$, then $\{V_\alpha\}$ is called an open cover of A. If every open cover has a finite subcover, then A is called a compact set. If X is a compact set, then X is called a compact topological space, or compact space for short.

Obviously, a closed subset of a compact set is also a compact set, and a single point set is always a compact set, but not necessarily a closed set.

Definition 2.1.3. Let $\{X, \mathscr{T}\}$ be a topological space and Y be a nonempty subset of X. Define $\mathscr{T}' = \{G \cap Y : G \in \mathscr{T}\}$. Then $(Y, \mathscr{T}')$ is a topological space and is called the subspace of $(X, \mathscr{T})$. We call $\mathscr{T}'$ the subtopology of $\mathscr{T}$.

Definition 2.1.4. Let X be a topological space. For any two different points $x \neq y$ in X, there exist open sets U, V with $x \in U$, $y \in V$, and $U \cap V = \emptyset$, then X is called a Hausdorff space or a T_2 space. The property described in the definition is called the second separation axiom, often referred to as the T_2 axiom.

By definition, in a Hausdorff space, a compact set must be a closed set. Of course, a single point set must be a closed set.

An important property about a Hausdorff space is that any pair of compact sets that do not intersect can be separated by a pair of open sets that do not intersect, that is, the following theorem holds.

Theorem 2.1.1. *Let X be a Hausdorff space, K and L be compact sets of X, and $K \cap L = \emptyset$. Then, there exist two open sets U and V, such that*

$$K \subset U, \quad L \subset V, \quad U \cap V = \emptyset.$$

Proof. We may assume that K and L are both nonempty (if $K = \emptyset$, then we can choose $U = \emptyset$, $V = X$ to obtain the conclusion). There are two situations to consider:

1. Let $K = \{x\}$ (i.e. K only contains one point). By the separation axiom of X, for any $y \in L$, there exist open sets U_y and V_y such that $x \in U_y$, $y \in V_y$ and $U_y \cap V_y = \emptyset$. Since L is a compact set, we can choose finite open sets V_{y_i}, $i = 1, \dots, m$ from $\{V_y : y \in L\}$ such that $\bigcup_{i=1}^m V_{y_i} \supset L$. Hence, we take $U = \bigcap_{i=1}^m U_{y_i}$ and $V = \bigcup_{i=1}^m V_{y_i}$, then U and V are desired.
2. Let K be a compact set. For every $x \in K$, according to (i), there exist open sets U_x and V_x such that

$$x \in U_x, \quad L \subset V_x, \quad U_x \cap V_x = \emptyset.$$

Since K is a compact set, we can choose finite open sets U_{x_j}, $j = 1, \dots, s$ from $\{U_x : x \in K\}$ such that $\bigcup_{j=1}^s U_{x_j} \supset K$. Hence, we take $U = \bigcup_{j=1}^s U_{x_j}$ and $V = \bigcap_{j=1}^s V_{x_j}$, then U and V are desired. $\square$

Corollary 2.1.2. *Let X be a Hausdorff space. If K is a compact set of X, U_1 and U_2 are open sets of X with $K \subset U_1 \cup U_2$, then there exist compact sets K_1 and K_2 such that $K_1 \subset U_1$, $K_2 \subset U_2$ and $K = K_1 \cup K_2$.*

Proof. Let $L_1 = K \setminus U_1$ and $L_2 = K \setminus U_2$. Obviously, L_1 and L_2 are both compact sets. Moreover, $L_1 \cap L_2 = \emptyset$. By Theorem 2.1.1, there exist V_1 and V_2 such that $L_1 \subset V_1$, $L_2 \subset V_2$ and $V_1 \cap V_2 = \emptyset$. Let $K_1 = K \setminus V_1$ and $K_2 = K \setminus V_2$. It can be seen that K_1 and K_2 are both compact sets. We have $K_1 \subset U_1$ since

$$K_1 = K \setminus V_1 = K \cap \complement V_1 \subset K \cap \complement L_1 = K \cap \complement(K \cap \complement U_1) = K \cap U_1.$$

Similarly, $K_2 \subset U_2$. Then we have

$$K = (K \setminus V_1) \cup (K \cap V_1) \subset (K \setminus V_1) \cup (K \cap \complement V_2) = K_1 \cup K_2,$$

which implies that $K = K_1 \cup K_2$. $\square$

Definition 2.1.5. Let X be a topological space. If any pair of disjoint closed sets of X can be separated by a pair of disjoint open sets, i.e. if A, B are closed sets and $A \cap B = \emptyset$, then there exist open sets U, V such that $A \subset U, B \subset V$ and $U \cap V = \emptyset$. Then X is called a normal topological space. This property is called the fourth separation axiom, often referred to as the T_4 axiom.

Corollary 2.1.3. *A compact Hausdorff space is normal.*

Proof. Since a closed set in a compact topological space is also a compact set, the conclusion is derived from Theorem 2.1.1. $\square$

Next, we introduce the notion of a continuous mapping on a topological space.

Definition 2.1.6. Let X and Y be topological spaces, a mapping $f : X \to Y$ is called continuous if for any open set $V \subset Y$, the preimage $f^{-1}(V)$ of V is an open set of X. Equivalently, for any closed set $F \subset Y$, $f^{-1}(F)$ is a closed set of X. When $Y = \mathbb{R}$ (or $\mathbb{C}$), the mapping is also called a real (or complex) function. We denote the set of continuous functions on X by $C(X)$.

Theorem 2.1.4 (Uryson's lemma). *Let X be a normal topological space. If E and F are closed sets of X with $E \cap F = \emptyset$, then there exists a continuous function $f : X \to [0, 1]$ such that*

$$f(x) = \begin{cases} 0, & x \in E, \\ 1, & x \in F. \end{cases}$$

Proof. Denote D as the set of all dyadic rational numbers in $(0, 1)$, i.e.

$$D = \left\{ \frac{m}{2^n} : m < 2^n; \ m, n \in \mathbb{N} \right\}.$$

By the normality of X, there exist open sets $U_{\frac{1}{2}}$ and $U'_{\frac{1}{2}}$ such that

$$E \subset U_{\frac{1}{2}}, \ F \subset U'_{\frac{1}{2}}, \ U_{\frac{1}{2}} \cap U'_{\frac{1}{2}} = \emptyset.$$

Obviously, $\overline{U_{\frac{1}{2}}} \cap U'_{\frac{1}{2}} = \emptyset$. Then,

$$E \subset U_{\frac{1}{2}} \subset \overline{U_{\frac{1}{2}}} \subset \complement F.$$

Using the normality of X for E and $\complement U_{\frac{1}{2}}$, there exist open sets $U_{\frac{1}{4}}$ and $U'_{\frac{1}{4}}$ such that

$$E \subset U_{\frac{1}{4}}, \quad \complement U_{\frac{1}{2}} \subset U'_{\frac{1}{4}}, \quad U_{\frac{1}{4}} \cap U'_{\frac{1}{4}} = \emptyset.$$

Then, we have

$$E \subset U_{\frac{1}{4}} \subset \overline{U_{\frac{1}{4}}} \subset \complement(\complement U_{\frac{1}{2}}) = U_{\frac{1}{2}}.$$

Using the normality of X for $\overline{U_{\frac{1}{2}}}$ and F, there exist open sets $U_{\frac{3}{4}}$ and $U'_{\frac{3}{4}}$ such that

$$\overline{U_{\frac{1}{2}}} \subset U_{\frac{3}{4}}, \quad F \subset U'_{\frac{3}{4}}, \quad U_{\frac{3}{4}} \cap U'_{\frac{3}{4}} = \emptyset.$$

Similarly, we have

$$\overline{U_{\frac{1}{2}}} \subset U_{\frac{3}{4}} \subset \overline{U_{\frac{3}{4}}} \subset \complement F.$$

Based on the above, we get

$$E \subset U_{\frac{1}{4}} \subset \overline{U_{\frac{1}{4}}} \subset U_{\frac{1}{2}} \subset \overline{U_{\frac{1}{2}}} \subset U_{\frac{3}{4}} \subset \overline{U_{\frac{3}{4}}} \subset \complement F.$$

By analogy, a sequence of open sets $\{U_r\}_{r \in D}$ is obtained, which satisfies

$$E \subset U_r \subset \overline{U_r} \subset U_s \subset \overline{U_s} \subset \complement F \quad (r, s \in D; \ r < s).$$

Define

$$f(x) = \begin{cases} 1, & x \notin \bigcup \{ U_r : r \in D \}, \\ \inf\{r : x \in U_r\}, & x \in \bigcup \{ U_r : r \in D \}, \end{cases}$$

then f satisfies

$$f(x) = \begin{cases} 0, & x \in E, \\ 1, & x \in F. \end{cases}$$

For the rest, it is only necessary to prove that f is continuous on X. By Definition 2.1.5 and the construction of open sets in $\mathbb{R}$, we only need to prove that for any $(\alpha, \beta) \subset \mathbb{R}$, $f^{-1}((\alpha, \beta))$ is an open set of X. In fact, it suffices to show that for any $(\alpha, \infty) \subset \mathbb{R}$, $f^{-1}((\alpha, \infty))$ is an open set of X and for any $(-\infty, \beta) \subset \mathbb{R}$, $f^{-1}((-\infty, \beta))$ is an open set of X. We only give the proof of the previous assertion.

Assume that $0 \leq \alpha < 1$. Considering the equivalent condition of $f(x) > \alpha$ is that there exists an $r \in D$, $r > \alpha$ such that $x \notin U_r$, and the fact that if $s \in D, r \in D, s < r$, then $\overline{U_s} \subset U_r$, the equivalent charaction of $f(x) > \alpha$ is that there exists $s \in D, s > \alpha$ such that $x \notin \overline{U_s}$. Thus, $f(x) > \alpha$ is equivalent to

$$x \in \bigcup \{\complement \overline{U_s} : s > \alpha, s \in D\}.$$

Therefore,

$$f^{-1}((\alpha, \infty)) = \bigcup \{\complement \overline{U_s} : s > \alpha, s \in D\}$$

is an open set of X. $\qquad\square$

Theorem 2.1.5 (Uryson's extension theorem). *Let X be a normal topological space, F be a closed subset of X, and $\varphi : F \to \mathbb{R}$ be a bounded continuous function. Then there exists a bounded continuous function $f : X \to \mathbb{R}$ such that for any $x \in F$, $f(x) = \varphi(x)$ and*

$$\sup \{|f(x)| : x \in X\} = \sup \{|\varphi(x)| : x \in F\}.$$

Proof. Write $\mu_0 = \sup \{|\varphi(x)| : x \in F\}$ and $\varphi_0 = \varphi$. Assume $\mu_0 > 0$. Define

$$A_0 = \left\{ x : x \in F, \; \varphi_0(x) \leq -\frac{\mu_0}{3} \right\},$$

$$B_0 = \left\{ x : x \in F, \; \varphi_0(x) \geq \frac{\mu_0}{3} \right\}.$$

Since $\varphi_0 \in C(F)$ and F is a closed set of X, then A_0, B_0 are closed sets of X and $A_0 \cap B_0 = \emptyset$. According to Theorem 2.1.4, there exists $f_0 \in C(X)$ such that

$$f_0(X) \subset \left[\frac{-\mu_0}{3}, \frac{\mu_0}{3}\right], \quad f_0(A_0) = \left\{\frac{-\mu_0}{3}\right\}, \quad f_0(B_0) = \left\{\frac{\mu_0}{3}\right\}.$$

Let $\varphi_1 = \varphi_0 - f_0$ on F, then $\varphi_1 \in C(F)$ and

$$\sup |\varphi_1|(F) = \mu_1 \le \frac{2}{3}\mu_0.$$

Define

$$A_1 = \left\{ x : x \in F, \ \varphi_1(x) \le -\frac{\mu_1}{3} \right\},$$

$$B_1 = \left\{ x : x \in F, \ \varphi_1(x) \ge \frac{\mu_1}{3} \right\}.$$

Take $f_1 \in C(X)$ such that

$$f_1(X) \subset \left[\frac{-\mu_1}{3}, \frac{\mu_1}{3} \right], \quad f_1(A_1) = \left\{ \frac{-\mu_1}{3} \right\}, \quad f_1(B_1) = \left\{ \frac{\mu_1}{3} \right\}.$$

Let $\varphi_2 = \varphi_1 - f_1$ on F, then $\varphi_2 \in C(F)$ and

$$\sup |\varphi_2|(F) = \mu_2 \le \frac{2}{3}\mu_1.$$

Repeat the above steps and we obtain a sequence of continuous functions $\{\varphi_n\}_{n=0}^{\infty}$ on F and a continuous function sequence $\{f_n\}_{n=0}^{\infty}$ on X which satisfy

$$\varphi_{n+1}(x) = \varphi_n(x) - f_n(x), \quad x \in F,$$

$$\mu_n = \sup |\varphi_n|(F),$$

$$\sup |f_n|(X) \le \frac{1}{3}\mu_n, \quad \mu_{n+1} \le \frac{2}{3}\mu_n$$

where $n = 0, 1, 2, \ldots$. Therefore,

$$\sup |\varphi_n|(F) \le \left(\frac{2}{3} \right)^n \mu_0, \quad \sup |f_n|(X) \le \frac{1}{3} \left(\frac{2}{3} \right)^n \mu_0.$$

Define

$$S_n(x) = \sum_{k}^{n} f_k(x), \quad n \in \mathbb{N}, \ x \in X,$$

then

$$|S_n(x) - S_{n+m}(x)| \le \sum_{k=1}^{m} |f_{n+k}(x)| \le \frac{1}{3} \left(\frac{2}{3} \right)^n \mu_0 \cdot \sum_{k=1}^{m} \left(\frac{2}{3} \right)^k,$$

which implies that $\{S_n\}$ converges uniformly on X. Let

$$f(x) = \lim_{n\to\infty} S_n(x),$$

then f is continuous on X and

$$\sup |f|(X) \le \sum_{n=0}^{\infty} \sup |f_n|(X) \le \frac{1}{3}\mu_0 \sum_{n=0}^{\infty} \left(\frac{2}{3}\right)^n = \mu_0.$$

For $x \in F$, we have

$$S_n(x) = \sum_{k=0}^{n} \left[\varphi_k(x) - \varphi_{k+1}(x)\right] = \varphi_0(x) - \varphi_{n+1}(x),$$

which shows that

$$f(x) = \lim_{n\to\infty} S_n(x) = \varphi_0(x) - \lim_{n\to\infty} \varphi_{n+1}(x) = \varphi_0(x). \qquad \square$$

2.2 Continuous Function on Locally Compact Hausdorff Space

Definition 2.2.1. A topological space X is called locally compact if for any $x \in X$, there exists an open set $U \ni x$ such that $\overline{U}$ is a compact set.

Obviously, $\mathbb{R}^n$ is a locally compact Hausdorff space (by Euclidean topology). In addition, a compact Hausdorff space is also locally compact. For simplicity, we write locally compact Hausdorff space as LCHS. LCHS has the following important properties.

Theorem 2.2.1. *If X is an LCHS, $K \subset X$ is compact, $U \subset X$ is open with $K \subset U$, then there exists an open set $V \subset X$ such that $\overline{V}$ is compact and $K \subset V \subset \overline{V} \subset U$.*

Proof.   We consider two cases.

(1) $K = \{x\}$ (i.e. K contains only one point). By Definition 2.2.1, there exists an open set $W \ni x$ such that $\overline{W}$ is a compact set. Assume $W \subset U$ (otherwise, replace W with $W \cap U$). Using Theorem 2.1.1 with $K = \{x\}$ and $L = \overline{W} \setminus W$, there exist open sets V_1

and V_2 such that $K \subset V_1$, $L \subset V_2$ and $V_1 \cap V_2 = \emptyset$. If $V = V_1 \cap W$ is taken, it is easy to prove that V satisfies the conclusion of theorem. In fact, the inclusion relation $\overline{V} \subset U$ can be derived from the following two series of relations:

$$V_1 \subset \complement V_2 \Rightarrow \overline{V}_1 \subset \complement V_2 \Rightarrow \overline{V} \subset \complement V_2 \subset \complement L,$$

$$\overline{V} \subset \overline{W} = L \cup W.$$

(2) K is a compact set of X. It can be known from (i) that for any $x \in K$, there exists an open set V_x such that $\overline{V}_x$ is a compact set and $\overline{V}_x \subset U$. By the compactness of K, a finite number of V_{x_i}, $i = 1, \ldots, m$, can be selected from $\{V_x : x \in K\}$ such that $\bigcup_{i=1}^{m} V_{x_i} \supset K$. Obviously, if $V = \bigcup_{i=1}^{m} V_{x_i}$ is taken, then V satisfies the conclusion of theorem. $\qquad\square$

Let X be an LCHS and denote

$$C_c(X) = \{f \in C(X) : \operatorname{supp} f \text{ is a compact set}\},$$

where $\operatorname{supp} f = \overline{\{x \in X : f(x) \neq 0\}}$ is called the support of f.

The following theorem is a generalization of the Uryson lemma in LCHS.

Theorem 2.2.2. *Let X be an LCHS, K be a compact set of X, U be an open set of X with $K \subset U$. Then there exists $f \in C_c(X)$ such that $\chi_K(x) \leq f(x) \leq \chi_U(x)$ and $\operatorname{supp} f \subset U$, where χ_E is the characteristic function of the set E.*

Proof. By Theorem 2.2.1, there exists an open set V such that $\overline{V}$ is compact and

$$K \subset V \subset \overline{V} \subset U.$$

By Corollary 2.1.3, $\overline{V}$ is a normal Hausdorff space (as a subspace of X). Therefore, by Theorem 2.1.4, there exists a continuous function $g : \overline{V} \to [0,1]$ on $\overline{V}$ such that

$$g(x) = \begin{cases} 1, & x \in K, \\ 0, & x \in \overline{V} \setminus V. \end{cases}$$

Now, define

$$f(x) = \begin{cases} g(x), & x \in \overline{V}, \\ 0, & x \in X \setminus \overline{V}. \end{cases}$$

It is easy to prove that f is a continuous function on X. In fact, for any closed set $E \subset [0,1]$, we have $f^{-1}(E) = g^{-1}(E)$ if $0 \notin E$ and $f^{-1}(E) = g^{-1}(E) \cup \complement V$ if $0 \in E$. Therefore, $f^{-1}(E)$ is closed for both cases. This implies that f is continuous on X. In addition, $\chi_K(x) \le f(x) \le \chi_U(x)$ is derived from $K \subset \operatorname{supp} f \subset \overline{V} \subset U$ and $0 \le f(x) \le 1$. $\square$

Theorem 2.2.3. *Let X be an LCHS, $f \in C_c(X)$, and $\operatorname{supp} f \subset \bigcup_{i=1}^m U_i$, where U_i is an open set of X. Then there exist $f_i \in C_c(X)$, $1 \le i \le m$ such that $f = \sum_{i=1}^m f_i$ and $\operatorname{supp} f_i \subset U_i$ $(1 \le i \le m)$. Moreover, we have $f_i \ge 0$ $(1 \le i \le m)$ if $f \ge 0$.*

Proof.   We may assume that $m = 2$. By Corollary 2.1.2, there exist compact sets K_1 and K_2 such that

$$K_1 \subset U_1, \ K_2 \subset U_2, \ \operatorname{supp} f = K_1 \cup K_2.$$

By Theorem 2.2.2, there exist $h_i \in C_c(X)$ $(i = 1, 2)$ such that

$$\chi_{K_i}(x) \le h_i(x) \le \chi_{U_i}(x), \quad \operatorname{supp} h_i \subset U_i \ (i = 1, 2).$$

Let

$$g_1 = h_1, \quad g_2 = h_2 - \min(h_1, h_2),$$

then $g_1 \ge 0$, $g_2 \ge 0$, $\operatorname{supp} g_1 \subset U_1$ and $\operatorname{supp} g_2 \subset U_2$. It is easy to prove

$$g_1 + g_2 = \max(h_1, h_2).$$

Therefore, $g_1(x) + g_2(x) = 1$ when $x \in \operatorname{supp} f = K_1 \cup K_2$. Take

$$f_1 = f g_1, \quad f_2 = f g_2,$$

then f_1 and f_2 are desired. $\square$

The following theorem is a generalization of Uryson extension theorem in LCHS.

Theorem 2.2.4 (Tietze's extension theorem). *Let* X *be an LCHS and* K *be a compact set of* X. *If* $f \in C(K)$, *then there exists* $\varphi \in C_c(X)$ *such that* $\varphi|_K = f$, *where* $\varphi|_K$ *denotes the restriction of* φ *on* K.

Proof. By Theorem 2.2.2, there exists $g \in C_c(X)$ such that $\chi_K \leq g \leq 1$. Write $F = \operatorname{supp} g$, then F is a compact set of X. As a subspace of X, F is compact and hence normal. Using Theorem 2.1.5 on the subspace F, we know that $h \in C(F)$ and $h|_K = f$. Define the function φ on X,

$$\varphi(x) = \begin{cases} g(x)h(x) = (gh)(x), & x \in F, \\ 0, & x \in X \setminus F, \end{cases}$$

then $\varphi|_K = f$ and $\operatorname{supp} \varphi \subset F$.

Let E be a closed set of $\mathbb{R}$. Obviously, $gh \in C(F)$ and $(gh)^{-1}(E)$ are closed subsets of F and thus closed subsets of X.

We consider two cases:

(i) $0 \notin E$: In this case, $\varphi^{-1}(E) = (gh)^{-1}(E)$ is a closed subset of F and X.

(ii) $0 \in E$: We have

$$\varphi^{-1}(E) = (gh)^{-1}(E) \cup \varphi^{-1}(\{0\}),$$

where $(gh)^{-1}(E)$ is a closed subset of F and

$$\varphi^{-1}(\{0\}) = \complement F \cup g^{-1}(\{0\}) \cup h^{-1}(\{0\}) = g^{-1}(\{0\}) \cup h^{-1}(\{0\}),$$

where $g^{-1}(\{0\})$ is a closed set of X, $h^{-1}(\{0\})$ is a closed set of F and thus a closed subset of X. Therefore, $\varphi^{-1}(\{0\})$ is a closed subset of X and then $\varphi^{-1}(E)$ is a closed set of X.

Hence, taking the above cases into account, we can obtain $\varphi \in C_c(X)$. $\qquad\square$

We introduce the concept of countable basis in the following.

Definition 2.2.2. Let $(X, \mathscr{T})$ be a topological space and $\mathscr{U} \subset \mathscr{T}$. For any $A \in \mathscr{T}$, if A can be expressed as the union of elements in $\mathscr{U}$, then $\mathscr{U}$ is called a basis of $(X, \mathscr{T})$. Moreover, if $\mathscr{U}$ is a countable set, then X is called a topological space with a countable basis $\mathscr{U}$.

Remark. We call a set whose cardinality does not exceed $\aleph_0$ countable set. That is, empty set, finite set and the set which is equal to the natural number set $\mathbb{N}$ are called countable sets.

Theorem 2.2.5. *Let X be an LCHS. If X has a countable basis $\mathscr{U}$, then every open set of X can be expressed as a union of countable compact sets.*

Proof. Let U be an open set of X and $x \in U$. According to Theorem 2.2.1, there exists an open set V such that $x \in V \subset \overline{V} \subset U$, and $\overline{V}$ is a compact set. By the definition of $\mathscr{U}$, there exists $U_x \in \mathscr{U}$ such that $x \in U_x \subset V$. Then $\overline{U_x}$ is also a compact set. Obviously, the set $\{U_x : x \in U\}$ is a subset of the countable set $\mathscr{U}$, and

$$U \subset \bigcup \left\{ \overline{U_x} : x \in U \right\} \subset U,$$

which implies that U is a union of countable compact sets.   □

Remark. In the following, if we say that X has a countable basis $\mathscr{U}$, then $X \in \mathscr{U}$ naturally. Otherwise, if X is added to $\mathscr{U}$, we can still obtain a countable basis.

Remark. Theorem 2.2.5 shows that under the condition of Theorem 2.2.5, any open set of X is an F_σ set.

2.3   Radon Measure and the Riesz Representation Theorem

Definition 2.3.1. Let X be a Hausdorff space. The σ-algebra generated by the open sets of X is called the Borel algebra on X and denoted by $\mathscr{B}(X)$. The element of $\mathscr{B}(X)$ is called the Borel set of X.

Definition 2.3.2. A measure on $\mathscr{B}(X)$ is called a Borel measure on X if it takes value on $[0, \infty)$ on every compact set.

Remark. Here, unlike [2], taking value on $[0, \infty)$ on every compact set is a precondition for a Borel measure.

Definition 2.3.3. Let X and Y be Hausdorff spaces. A mapping $f : X \to Y$ is called Borel measurable, which means that for each $B \in \mathscr{B}(Y)$, there is $f^{-1}(B) \in \mathscr{B}(X)$.

The following theorem shows that continuous functions on Hausdorff space must be Borel measurable. However, Borel measurable functions are not necessarily continuous.

Theorem 2.3.1. *Let X and Y be Hausdorff spaces. If a mapping $f : X \to Y$ is continuous, then f is Borel measurable.*

Proof. By Definition 2.1.5, when U is an open set of Y, $f^{-1}(U)$ is an open set of X, and thus, $f^{-1}(U) \in \mathscr{B}(X)$. Denote $\mathscr{F} = \{B \subset Y : f^{-1}(B) \in \mathscr{B}(X)\}$. Then $\mathscr{F}$ contains all open sets of Y. It is easy to verify that $\mathscr{F}$ is a σ-algebra on Y. Then we have $\mathscr{F} \supset \mathscr{B}(Y)$, which implies that for every $B \in \mathscr{B}(Y)$, $f^{-1}(B) \in \mathscr{B}(X)$. □

Definition 2.3.4. Let $\mathscr{A}$ be a σ-algebra on a Hausdorff space X and $\mathscr{B}(X) \subset \mathscr{A}$. A measure μ on $\mathscr{A}$ is called regular if μ has the following properties:

(a) for any compact set $K \subset X$, $\mu(K) < \infty$;
(b) for any $A \in \mathscr{A}$, $\mu(A) = \inf\{\mu(U) : U$ is an open set, $U \supset A\}$;
(c) for any open set $U \subset X$, $\mu(U) = \sup\{\mu(K) : K$ is a compact set, $K \subset U\}$.

Remark. The property (b) is called outer regularity condition and (c) is called inner regularity condition.

Definition 2.3.5. The regular measure on $\mathscr{B}(X)$ is called a regular Borel measure on X or a Radon measure on X.

Definition 2.3.4 gives the conditions for the Borel measure on LCHS to be a regular measure (i.e. Radon measure). If the discussion is limited to an LCHS with a countable basis, we will prove that the Borel measure on it is always regular below.
Indeed, we have the following theorem.

Theorem 2.3.2. *Let X be an LCHS with a countable basis and μ be a Borel measure on X. Then every Borel set is regular with respect to μ, that is, the conditions (b) and (c) of Definition 2.3.4 hold.*

Proof. From Theorem 2.2.5, X can be expressed as $X = \bigcup_i K_i$, where K_i is compact. Then by Theorem 2.2.1, there exists an open set V_i of X such that $\overline{V_i}$ is compact and

$$K_i \subset V_i \subset \overline{V_i} \subset X.$$

The above fact indicates that $X = \bigcup_i V_i$ and $\mu(V_i) < \infty$.

Let $A \in \mathscr{B}(X)$. For every $n \in \mathbb{N}$, we denote that $\mu_n(A) = \mu(A \cap V_n)$. Obviously, $\mu_n(A) < \infty$, so μ_n is a finite Borel measure. We first prove that every Borel set with respect to μ_n satisfies (b) and (c). Then the conclusion of Theorem 2.3.2 is proved by transiting from μ_n to μ.

Fix $n \in \mathbb{N}$ and write $\nu = \mu_n$. Denote

$$\mathscr{A} = \{A \in \mathscr{B}(X) : A \text{ is regular}\},$$

where A is regular, which means that A satisfies Definition 2.3.4(b) and (c) with respect to ν. Obviously, every open set U belongs to $\mathscr{A}$. According to Theorem 2.2.5, U can be expressed as a countable union of compact sets. In the following, we verify that $\mathscr{A}$ is a σ-algebra, and then $\mathscr{A} = \mathscr{B}(X)$.

We first prove that $\mathscr{A}$ is closed under finite complements. Let $A \in \mathscr{A}$. For any $\varepsilon > 0$, there exist an open set $G \supset A$ and a compact set $K \subset A$ such that $\nu(G \setminus K) < \varepsilon$. Then

$$\nu(\complement K \setminus \complement G) = \mu(\complement K \cap V_n \setminus \complement G \cap \overline{V_n}) < \varepsilon,$$

the set $\complement K$ is open, $\complement K \supset \complement A$, $\complement G \cap \overline{V_n}$ are compact and $(\complement G \cap \overline{V_n}) \subset \complement A$. This proves that $\complement A \in \mathscr{A}$.

Then we prove that $\mathscr{A}$ is closed under countable unions. Let $A_k \in \mathscr{A}$, $k \in \mathbb{N}$. For any $\varepsilon > 0$, there exist an open set $G_k \supset A_k$ and a compact set $K_k \subset A_k$ such that

$$\nu(G_k \setminus K_k) < \varepsilon 2^{-k}.$$

Write $A = \bigcup_{k=1}^{\infty} A_k$, $G = \bigcup_{k=1}^{\infty} G_k$, and $F = \bigcup_{k=1}^{\infty} K_k$, we have $G \supset A \supset F$ and

$$\nu(G \setminus F) \leq \sum_{k=1}^{\infty} \nu(G_k \setminus K_k) \leq \varepsilon.$$

The above inequality indicates $A \in \mathscr{A}$, since $\lim_{N \to \infty} \nu\left(\bigcup_{i=1}^{N} K_i\right) = \nu(F)$ and $\bigcup_{i=1}^{N} K_i$ is a compact set contained in A.

Now, consider the measure μ. Let $E \in \mathscr{B}(X)$. Then for any $\varepsilon > 0$ and $n \in \mathbb{N}$, there exists an open set $G_n \supset E$ and a compact set $K_n \subset E$ such that $\mu_n(G_n \setminus K_n) < \varepsilon 2^{-n}$. Let $G = \bigcup_{n=1}^{\infty}(G_n \cap V_n) \supset \bigcup_{n=1}^{\infty}(E \cap V_n) = E$. Obviously, G is an open set. Let $F = \bigcup_{n=1}^{\infty} K_n$, then $F \subset E$ and

$$\mu(G \setminus F) \leq \sum_{n=1}^{\infty} \mu(G_n \cap V_n \setminus K_n) = \sum_{n=1}^{\infty} \mu_n(G_n \setminus K_n) < \varepsilon.$$

By this estimate and the fact that F is a countable union of compact sets, we can obtain that either $\mu(E) < \infty$ or $\mu(E) = \infty$, E is always regular with respect to μ. $\qquad\square$

Next, let X be an LCHS. We study the relationship between the Radon measures on X and the linear functionals on $C_c(X)$.

Let μ be a Radon measure, then the mapping L :

$$f \mapsto \int_X f \, d\mu$$

defines a linear functional on $C_c(X)$. Obviously, if $f \in C_c(X)$ and $f \geq 0$, then $L(f) \geq 0$.

Definition 2.3.6. Let X be an LCHS. A linear functional I on $C_c(X)$ is called positive if $I(f) \geq 0$ for any $f \in C_c(X)$ and $f \geq 0$.

Then we see that every Radon measure μ corresponds to a positive linear functional L such that for any $f \in C_c(X)$, $L(f) = \int_X f \, d\mu$.

We are concerned with the reverse problems:

(1) Whether a positive linear functional on $C_c(X)$ must have the form $f \mapsto \int_X f \, d\mu$, where μ is a Radon measure on X?
(2) Is it unique if the Radon measure μ exists?

The following Riesz representation theorem will answer these questions.

Definition 2.3.7. The symbol $f \prec U$ denotes $0 \leq f \leq 1$, $f \in C_c(X)$ and $\operatorname{supp} f \subset U$. Specially, we stipulate $0 \prec \emptyset$.

Lemma 2.3.3. *Let X be an LCHS and μ be a Radon measure on X. If U is an open set of X, then*

$$\mu(U) = \sup \left\{ \int_X f \, d\mu : f \in C_c(X), \; f \prec U \right\}.$$

Proof. By inequality

$$\int_X f \, d\mu \leq \mu(\text{supp} f) \leq \mu(U),$$

it suffices to prove

$$\mu(U) \leq \sup\left\{\int_X f \, d\mu : f \in C_c(X), \ f \prec U\right\}.$$

Let $\alpha < \mu(U)$. By the regularity condition (c) of μ, there exists a compact set $K \subset U$ such that $\alpha < \mu(K)$. From Theorem 2.2.2, there exists $f \in C_c(X)$ such that $\chi_K \leq f$ and $f \prec U$. Then

$$\alpha < \mu(K) = \int_X \chi_K \, d\mu \leq \int_X f \, d\mu.$$

Therefore,

$$\alpha < \sup\left\{\int_X f \, d\mu : f \in C_c(X), \ f \prec U\right\}.$$

Take $\alpha \to \mu(U)$ and the inequality to be proved is obtained. $\qquad\square$

Lemma 2.3.4. *Let X be an LCHS. I is a linear functional on $C_c(X)$. For the subsets of X, we define the set function μ^* as follows:*

(α) *When U is an open set of X, $\mu^*(U) = \sup\{I(f) : f \in C_c(X), f \prec U\}$.*
(β) *For any $A \subset X$, $\mu^*(A) = \inf\{\mu^*(U) : U \text{ is open, } U \supset A\}$.*

Then, μ^* is an outer measure on X and every Borel subset of X is μ^*-measurable.

Proof. In order to prove that μ^* is an outer measure on X, we only need to prove the countable subadditivity of μ^*:

$$\mu^*\left(\bigcup_i A_i\right) \leq \sum_{i=1}^{\infty} \mu^*(A_i).$$

We first prove that the above inequality holds for any open set sequence $\{U_i\}$. Let $f \in C_c(X)$ and $f \prec \bigcup_i U_i$, then $\bigcup_i U_i$ is an open cover of the compact set supp f. Then, there exists $N \in \mathbb{N}$ such that

$$\text{supp}\, f \subset \bigcup_{i=1}^{N} U_i.$$

By Theorem 2.2.3, we have

$$f = \sum_{i=1}^{N} f_i, \quad f_i \in C_c(X), \quad f_i \prec U_i, \quad 1 \le i \le N.$$

Then,

$$I(f) = \sum_{i=1}^{N} I(f_i) \le \sum_{i=1}^{N} \mu^*(U_i) \le \sum_{i=1}^{\infty} \mu^*(U_i).$$

Taking the supremum over $\{f : f \in C_c(X),\ f \prec \bigcup_i U_i\}$ at both sides of the above inequalities and by (α), we can conclude that

$$\mu^*\left(\bigcup_i U_i\right) \le \sum_i \mu^*(U_i).$$

Second, in order to prove that μ^* also has countable subadditivity for any subset $\{A_i\}$ of X, we may assume

$$\sum_{i=1}^{\infty} \mu^*(A_i) < \infty,$$

then for every $i \in \mathbb{N}$, there exists an open set $U_i \supset A_i$ such that

$$\mu^*(U_i) \le \mu^*(A_i) + \varepsilon 2^{-i},$$

thus

$$\mu^*\left(\bigcup_i A_i\right) \le \mu^*\left(\bigcup_i U_i\right) \le \sum_{i=1}^{\infty} \mu^*(U_i) \le \sum_{i=1}^{\infty} \mu^*(A_i) + \varepsilon.$$

Therefore, the above assertion holds, that is, μ^* is an outer measure.

In order to prove that any Borel subset is μ^*-measurable, by Theorem 1.6.1 in Chapter 1, it suffices to prove that any open set U of X is μ^*-measurable, that is, for any $A \subset X$,

$$\mu^*(A) = \mu^*(A \cap U) + \mu^*(A \cap \complement U).$$

In fact, we just need to prove that for any $A \subset X$,

$$\mu^*(A) \geq \mu^*(A \cap U) + \mu^*(A \cap \complement U). \qquad (*)$$

Without loss of generality, we assume $\mu^*(A) < \infty$. By (β), there exists an open set $V \supset A$ satisfying

$$\mu^*(V) \leq \mu^*(A) + \varepsilon.$$

If we prove

$$\mu^*(V) \geq \mu^*(V \cap U) + \mu^*(V \cap \complement U) - 2\varepsilon, \qquad (**)$$

then combining the above two inequalities, we get

$$\mu^*(A) + \varepsilon \geq \mu^*(A \cap U) + \mu^*(A \cap \complement U) - 2\varepsilon,$$

then $(*)$ holds since ε is arbitrary. Therefore, we just need to prove $(**)$.

By (α), there exists $f_1 \in C_c(X)$ satisfying

$$I(f_1) \geq \mu^*(V \cap U) - \varepsilon,$$

and $f_1 \prec V \cap U$. Write $K = \operatorname{supp} f_1$, then the open set $V \cap \complement K \supset V \cap \complement U$. By (α), there exists $f_2 \in C_c(X)$ satisfying

$$I(f_2) \geq \mu^*(V \cap \complement K) - \varepsilon \geq \mu^*(V \cap \complement U) - \varepsilon,$$

and $f_2 \prec V \cap \complement K$. Obviously, $f_1 + f_2 \prec V$ and $I(f_1 + f_2) \leq \mu^*(V)$. The latter is

$$[\mu^*(V \cap U) - \varepsilon] + [\mu^*(V \cap \complement U) - \varepsilon] \leq \mu^*(V),$$

which is $(**)$. $\qquad\qquad \square$

Lemma 2.3.5. *Let X, I, μ^* be as described in Lemma* 2.3.4, $A \subset X$ *and $f \in C_c(X)$, then*

(1) *if $\chi_A \leq f$, then $\mu^*(A) \leq I(f)$;*
(2) *if $0 \leq f \leq \chi_A$ and A is compact, then $I(f) \leq \mu^*(A)$.*

Proof. (1) Take $0 < \varepsilon < 1$ and write

$$U_\varepsilon = \{x \in X : f(x) > 1 - \varepsilon\}.$$

Obviously, U_ε is an open set and $U_\varepsilon \supset A$. By (α), we have

$$\mu^*(U_\varepsilon) = \sup\{I(g) : g \prec U_\varepsilon\}.$$

Then,

$$g \leq \chi_{U_\varepsilon} \leq \left(\frac{1}{1 - \varepsilon}\right) f,$$

which follows that

$$\mu^*(U_\varepsilon) \leq \left(\frac{1}{1 - \varepsilon}\right) I(f),$$

so

$$\mu^*(A) \leq \mu^*(U_\varepsilon) \leq \left(\frac{1}{1 - \varepsilon}\right) I(f).$$

From the above inequalities and the arbitrariness of ε, the conclusion of (1) is obtained.

(2) Let U be an open set and $U \supset A$. Obviously, $f \prec U$. Then

$$I(f) \leq \mu^*(U).$$

Taking the infimum over the open set $U \supset A$ at both sides of the above inequality and by (β), we get $I(f) \leq \mu^*(A)$. $\square$

Lemma 2.3.6. *Let X, I, μ^* be as described in Lemma* 2.3.4. *If μ is the restriction of μ^* on $\mathscr{B}(X)$, then μ is a Radon measure and for any $f \in C_c(X)$,*

$$I(f) = \int_X f d\mu.$$

Proof.   By Theorem 2.2.2, for any compact set K, there exists an $f \in C_c(X)$ such that $\chi_K \leq f$ and by Lemma 2.3.5, $\mu(K) < \infty$. Therefore, the measure μ satisfies condition (a).

Let $A \in \mathscr{B}(X)$. By (β),

$$\mu(A) = \mu^*(A) = \inf\{\mu^*(U) : U \text{ is an open set}, \ U \supset A\}$$
$$= \inf\{\mu(U) : U \text{ is an open set}, \ U \supset A\}.$$

Thus, μ satisfies the outer regularity condition (b).

Let U be an open set. From (α) and Lemma 2.3.5 (2), we have

$$\mu(U) = \mu^*(U) = \sup\{I(f) : f \prec U\}$$
$$\leq \sup\{\mu^*(\operatorname{supp} f) : f \prec U\}$$
$$= \sup\{\mu(\operatorname{supp} f) : f \prec U\}$$
$$\leq \sup\{\mu(K) : K \text{ is a compact set}, \ K \subset U\}.$$

Combining the above inequality with

$$\mu(U) \geq \sup\{\mu(K) : K \text{ is a compect set}, \ K \subset U\},$$

we obtain that μ satisfies the inner regularity condition (c).

Therefore, μ is a Radon measure on X. Next, we establish the equation

$$I(f) = \int_X f \, d\mu.$$

Obviously, we only need to consider the case of $f \geq 0$.

For any $\varepsilon > 0$ and any $n \in \mathbb{N}$, define

$$f_n(x) = \begin{cases} 0, & f(x) \leq (n-1)\varepsilon, \\ f(x) - (n-1)\varepsilon, & (n-1)\varepsilon < f(x) < n\varepsilon, \\ \varepsilon, & f(x) \geq n\varepsilon. \end{cases}$$

Then, $f_n \in C_c(X)$ and $\operatorname{supp} f_n = \overline{\{x : f(x) > (n-1)\varepsilon\}}$ is also a compact set as a closed subset of $\operatorname{supp} f$. Since f is a bounded function, there exists an $N \in \mathbb{N}$ such that when $n > N$, $f_n = 0$. Therefore,

$$f = \sum_{n=1}^{N} f_n.$$

Write

$$K_0 = \operatorname{supp} f, \quad K_n = \{x \in X : f(x) \geq n\varepsilon\}, \quad n \in \mathbb{N}.$$

It is easy to verify that $\varepsilon \chi_{K_n} \leq f_n \leq \varepsilon \chi_{K_{n-1}}$ and $\varepsilon \mu(K_n) \leq \int_X f_n \, d\mu \leq \varepsilon \mu(K_{n-1})$ $(1 \leq n \leq N)$. If we rewrite the inequality $\varepsilon \chi_{K_n} \leq f_n \leq \varepsilon \chi_{K_{n-1}}$ as $\chi_{K_n} \leq \frac{1}{\varepsilon} f_n \leq \chi_{K_{n-1}}$, by Lemma 2.3.5,

$$\varepsilon \mu(K_n) \leq I(f_n) \leq \varepsilon \mu(K_{n-1}) \quad (1 \leq n \leq N).$$

Then

$$\varepsilon \sum_{n=1}^{N} \mu(K_n) \leq \int_X f \, d\mu \leq \varepsilon \sum_{n=0}^{N-1} \mu(K_n)$$

and

$$\varepsilon \sum_{n=1}^{N} \mu(K_n) \leq I(f) \leq \varepsilon \sum_{n=0}^{N-1} \mu(K_n).$$

Therefore,

$$\left| I(f) - \int_X f \, d\mu \right| \leq \varepsilon \mu(\operatorname{supp} f).$$

Since ε is arbitrary, it follows that $I(f) = \int_X f \, d\mu$. $\qquad \square$

Theorem 2.3.7 (Riesz representation theorem). *Let X be an LCHS. If I is a positive linear functional on $C_c(X)$, then there exists a unique Radon measure μ on X such that for any $f \in C_c(X)$, $I(f) = \int_X f \, d\mu$.*

Proof. The existence of μ is obtained by Lemma 2.3.6. Now, we prove the uniqueness. Suppose that there exist two Radon measures μ and ν such that for any $f \in C_c(X)$,

$$I(f) = \int_X f \, d\mu = \int_X f \, d\nu,$$

then by Lemma 2.3.3, we know that for any open set $U \subset X$, $\mu(U) = \nu(U)$. From the outer regularity condition (b) of μ and ν, for any $A \in$

$\mathscr{B}(X)$, we have $\mu(A) = \nu(A)$. Therefore, the Riesz representation theorem is proved. □

2.4   The Luzin Theorem

Lemma 2.4.1. *Let X be an LCHS and μ be a Radon measure on X. If*

$$E \in \mathscr{B}(X) \ and \ \mu(E) < \infty,$$

then

$$\mu(E) = \sup\{\mu(K) : K \subset E, \ K \ is \ a \ compact \ set\}.$$

Proof.   For any $\varepsilon > 0$, by the regularity of μ, there exists an open set $U \supset E$ such that

$$\mu(U) < \mu(E) + \varepsilon,$$

and there exists a compact set $L \subset U$ such that

$$\mu(L) > \mu(U) - \varepsilon.$$

By $\mu(U \setminus E) < \varepsilon$ and the outer regularity of μ, there exists an open set $V \supset U \setminus E$, which satisfies $\mu(V) < \varepsilon$. Let $K = L \setminus V$, then K is a compact set and

$$\mu(K) = \mu(L) - \mu(L \cap V) > \mu(U) - \varepsilon - \mu(V) > \mu(E) - 2\varepsilon.$$

We obtain the required conclusion. □

Theorem 2.4.2. *Let X be an LCHS and μ be a Radon measure on X. If $1 \le p < \infty$, then $C_c(X)$ is dense in $L^p(\mu)$, where $L^p(\mu)$ denotes the space $L^p(X, \mathscr{B}(X), \mu)$.*

Proof.   Let $f \in L^p(\mu)$, $f \ge 0$. Then, there exists a sequence of simple functions $\varphi_n, n \in \mathbb{N}$ such that for any $x \in X$,

$$0 \le \varphi_n \le \varphi_{n+1} \le f, \ \lim_{n \to \infty} \varphi_n(x) = f(x).$$

Then,

$$0 \le f - \varphi_n \le f.$$

According to the Lebesgue dominated convergence theorem, we obtain

$$\lim_{n \to \infty} \|f - \varphi_n\|_p = 0.$$

It can be seen that the simple function class is dense in $L^p(\mu)$. Therefore, it suffices to prove that a characteristic function χ_E can be approximate by the functions in $C_c(X)$ (in the sense of L^p norm), where $E \in \mathscr{B}(X)$ and $\mu(E) < \infty$. By Lemma 2.4.1, there exists a compact set $K \subset E$ such that

$$\mu(K) > \mu(E) - \frac{\varepsilon}{2}.$$

By the outer regularity of μ, there exists an open set $U \supset E$ such that

$$\mu(U) < \mu(E) + \frac{\varepsilon}{2}.$$

Therefore,

$$\mu(U \setminus K) < \varepsilon.$$

Then, by Theorem 2.2.2, there exists an $f \in C_c(X)$ such that $\chi_K \leq f \leq \chi_U$. Therefore, we have

$$\|\chi_E - f\|_p^p = \int_{U \setminus K} |\chi_E - f|^p d\mu \leq \mu(U \setminus K) < \varepsilon,$$

which confirms the previous assertion. $\qquad\Box$

Theorem 2.4.3 (Luzin's theorem). *Let X be an LCHS, μ be a Radon measure on X and the function $f : X \to \mathbb{R}$ be Borel measurable taking finite value almost everywhere. Let $E = \{x \in X : f(x) \neq 0\}$. If $\mu(E) < \infty$, then for any $\varepsilon > 0$, there exists $g \in C_c(X)$ such that*

$$\mu(\{x \in X : f(x) \neq g(x)\}) < \varepsilon.$$

Proof.   We consider the following two cases.

(i) Let f be a bounded measurable function. Then, $f \in L^1(\mu)$. By Theorem 2.4.2, there exists a sequence of functions $g_n \in C_c(X)$ such that

$$\|g_n - f\|_1 \to 0 \quad (n \to \infty).$$

Therefore, there exists a subsequence of $\{g_n\}$ (still denoted by $\{g_n\}$) such that

$$g_n \to f \quad \text{a.e.}$$

By $\mu(E) < \infty$ and the Egorov theorem (Theorem 1.2.3 in Chapter 1), there exists a set $A \subset E$ such that $\mu(E \setminus A) < \varepsilon$ and $\{g_n\}$ converges uniformly to f on A.

By the outer regularity of μ, there exists an open set $U \supset E$ such that

$$\mu(U \setminus E) < \varepsilon.$$

It follows from Lemma 2.4.1 that there exists a compact set $K \subset A$ such that

$$\mu(A \setminus K) < \varepsilon.$$

Since $\{g_n\}$ converges uniformly to f on $K \subset A$, the restriction $f|_K$ of f on K is a continuous function. Thus, by the Tietze extension theorem (Theorem 2.2.4), there exists $g \in C_c(X)$ such that $g(x) = f(x)$ and $\operatorname{supp} g \subset U$ for $x \in K$. However,

$$\{x \in X : f(x) \neq g(x)\} \in U \setminus K,$$

therefore

$$\mu(\{x \in X : f(x) \neq g(x)\}) \leq \mu(U \setminus K)$$
$$\leq \mu(U \setminus E) + \mu(E \setminus A) + \mu(A \setminus K) < 3\varepsilon.$$

Thus, the conclusion of the theorem holds for bounded measurable functions.

(ii) Let f be an unbounded measurable function. We assume that $E_n = \{x \in E : |f(x)| \leq n\}$, then $E_n \nearrow E$ $(n \to \infty)$. Therefore, there exists N such that for $n > N$,

$$\mu(E \setminus E_n) < \varepsilon.$$

Consider a fixed $n > N$. Using (i) with the function $f\chi_{E_n}$, there exists a $g \in C_c(X)$ such that

$$\mu(\{x \in X : g(x) \neq f(x)\chi_{E_n}(x)\}) < \varepsilon.$$

Therefore,

$$\mu(\{x : g \neq f\}) \leq \mu(\{x : g \neq f\chi_{E_n}\}) + \mu(\{x : f\chi_{E_n} \neq f\})$$

$$\leq \mu(\{x : g \neq f\chi_{E_n}\}) + \mu(E \setminus E_n)$$

$$< 2\varepsilon. \qquad \Box$$

2.5 Radon Product of Measures (Regular Product)

Let $(X, \mathscr{A}, \mu)$ and $(Y, \mathscr{B}, \nu)$ be two measure spaces. In Section 1.7, we have discussed their product spaces. Recall Definition 1.7.3, for $A \in \mathscr{A}, B \in \mathscr{B}$, we call $A \times B$ a measurable rectangle and $\mu(A)\nu(B)$ as area. Based on the area of the measurable rectangle, the outer measure λ^* is defined on $X \times Y$. Then, according to the Carathéodory condition, we determine a σ-algebra $\mathscr{M}$, and the restriction of λ^* on $\mathscr{M}$ is denoted by $\mu \times \nu$, which is called the product of measure μ and ν. We know that $\mathscr{M} \supset \mathscr{A} \times \mathscr{B}$, where $\mathscr{A} \times \mathscr{B}$ denotes the σ-algebra generated by measurable rectangles. Definition 1.7.3 says that the measure spaces $(X \times Y, \mathscr{M}, \mu \times \nu)$ and $(X \times Y, \mathscr{A} \times \mathscr{B}, \mu \times \nu)$ are called the product spaces of $(X, \mathscr{A}, \mu)$ and $(Y, \mathscr{B}, \nu)$. We know that the product measure $\mu \times \nu$ satisfies the condition for any $A \in \mathscr{A}$, $B \in \mathscr{B}$ with $(\mu \times \nu)(A \times B) = \mu(A)\nu(B)$.

Now, suppose X and Y are LCHS and μ, ν are, respectively, the Radon measure on X and Y. We will naturally ask whether the product measure $\mu \times \nu$ is the Radon measure on $X \times Y$.

We first recall the concept of product topology (see [11], p. 114). Let $\mathscr{U}, \mathscr{V}$ be the topology basis of X and Y, respectively. Then, denote $\mathscr{W} = \mathscr{U} \times \mathscr{V}$ as the topology bases of $X \times Y$. When discussing the product space, we only consider the product topology defined in this way. If X and Y are LCHS, then the product space $X \times Y$ is also LCHS (see [11], p. 192). If X and Y have a countable basis, so does $X \times Y$.

Obviously, the σ-algebra $\mathscr{M}$ in the product measure space $(X \times Y, \mathscr{M}, \mu \times \nu)$ must contain the σ-algebra $\mathscr{A} \times \mathscr{B}$, and for any $A \in$

$\mathscr{B}(X)$ and $B \in \mathscr{B}(Y)$,

$$(\mu \times \nu)(A \times B) = \mu(A)\nu(B). \tag{1}$$

It is natural to ask the following questions:

(1) What is the relationship between the σ-algebra $\mathscr{M}$ in the product measure space $(X \times Y, \mathscr{M}, \mu \times \nu)$ and the Borel algebra $\mathscr{B}(X \times Y)$? If $\mathscr{M} \supset \mathscr{B}(X \times Y)$, is $\mu \times \nu$ (limited to $\mathscr{B}(X \times Y)$) a Radon measure?

(2) If the relationship between $\mathscr{M}$ and $\mathscr{B}(X \times Y)$ is unknown, is there any way to define a Radon measure γ on $X \times Y$ such that it satisfies condition (1)? If such γ exists, is it unique?

When X, Y have a countable base, Question 1 has a positive answer. For Question 2, we will use the Riesz representation theorem to give the answer.

Lemma 2.5.1. (a) *If X and Y are LCHS, then*

$$\mathscr{B}(X) \times \mathscr{B}(Y) \subset \mathscr{B}(X \times Y);$$

(b) *if X and Y are LCHS with countable bases, then*

$$\mathscr{B}(X) \times \mathscr{B}(Y) = \mathscr{B}(X \times Y).$$

Proof.   By the definition of product topology, we have

$$A \times B = (A \times Y) \cap (X \times B), \quad A \in \mathscr{B}(X), \ B \in \mathscr{B}(Y).$$

Considering a continuous mapping $\pi : X \times Y \to X$ such that $\pi((x, y)) = x$ and Theorem 2.3.1, it is known that π is Borel measurable, that is, when $A \in \mathscr{B}(X)$,

$$\pi^{-1}(A) = A \times Y \in \mathscr{B}(X \times Y).$$

Similarly, it can be proved that when $B \in \mathscr{B}(Y)$, there is $X \times B \in \mathscr{B}(X \times Y)$. So, $A \times B \in \mathscr{B}(X \times Y)$. It can be seen that (a) is established.

For the proof of (b), let $\mathscr{U}$ and $\mathscr{V}$ be countable bases of X and Y, respectively. Then $\mathscr{W} = \{U \times V : U \in \mathscr{U}, \ V \in \mathscr{V}\}$ is a countable base of $X \times Y$. Therefore, any open set in $X \times Y$ can be expressed as the union of countable sets in $\mathscr{W}$, and each set in $\mathscr{W}$ belongs to

$\mathscr{B}(X) \times \mathscr{B}(Y)$. Thus, any open set in $X \times Y$ belongs to $\mathscr{B}(X) \times \mathscr{B}(Y)$, from which $\mathscr{B}(X \times Y) \subset \mathscr{B}(X) \times \mathscr{B}(Y)$ is derived. $\qquad \square$

The following theorem partially answers the first question.

Theorem 2.5.2. *Suppose that X and Y are LCHS with countable bases, and μ and ν are Radon measures on X and Y, respectively. Then the product measure $\mu \times \nu$ is a Radon measure on $X \times Y$.*

Proof. By Lemma 2.5.1(b), we have

$$\mathscr{B}(X) \times \mathscr{B}(Y) = \mathscr{B}(X \times Y).$$

Therefore, we only need to prove that $\mu \times \nu$ is a Radon measure. In fact, by Theorem 2.3.2, it suffices to prove $\mu \times \nu$ is finite on every compact set of $X \times Y$.

Let K be a compact set of $X \times Y$. Consider continuous mappings

$$\pi_1 : X \times Y \to X, \ \pi_1(x, y) = x,$$

$$\pi_2 : X \times Y \to Y, \ \pi_2(x, y) = y.$$

Write $K_1 = \pi_1(K)$ and $K_2 = \pi_2(K)$. Obviously, $K \subset K_1 \times K_2$, where K_1 and K_2 are compact sets of X and Y, respectively, so

$$(\mu \times \nu)(K) \leq (\mu \times \nu)(K_1 \times K_2) = \mu(K_1)\nu(K_2) < \infty. \qquad \square$$

To answer the second question, we need two steps:

(i) Define an appropriate Randon measure γ on the space $X \times Y$ such that it satisfies the equality (1).
(ii) Prove the uniqueness of γ.

In order to determine the Radon measure γ on $X \times Y$, we will use the Riesz representation theorem.

Lemma 2.5.3. *Suppose that X and Y are LCHS with countable bases, μ and ν are Radon measures on X and Y, respectively. If $f \in C_c(X \times Y)$, then*

(a) *for each $x \in X$, $f(x, \cdot) \in C_c(Y)$;*
(b) *define $\varphi(x) = \int_Y f(x, y)$, then $\varphi \in C_c(X)$.*

Proof. Write $K = \operatorname{supp} f(x,y)$, then K is a compact set of $X \times Y$. Denote

$$K_1 = \{x \in X : (x,y) \in K\}, \quad K_2 = \{y \in Y : (x,y) \in K\} \text{ and}$$

it is easy to see that K_1 is a compact set of X and K_2 is a compact set of Y. For each $x \in X$, $f(x,\cdot)$ is a continuous function on Y and $\operatorname{supp} f(x,\cdot) \subset K_2$. So, $f(x,\cdot) \in C_c(Y)$. Therefore, (a) is established.

Fix $x_0 \in X$. For each $y \in K_2$, by the continuity of f at (x_0, y), there exists an open neighborhood U_y of x_0 and an open neighborhood V_y of y such that for any $x \in U_y$, $y' \in V_y$,

$$|f(x,y') - f(x_0,y)| < \frac{\varepsilon}{2}.$$

Then $|f(x,y') - f(x_0,y')| < \varepsilon$. Obviously, $\{V_y : y \in K_2\}$ is an open cover of K_2, so there exist finite number of V_{y_i}, $i = 1, \ldots, m$, such that

$$\bigcup_{i=1}^{m} V_{y_i} \supset K_2.$$

Take $U = \bigcap_{i=1}^{m} U_{y_i}$, then when $x \in U$, $y \in K_2$,

$$|f(x,y) - f(x_0,y)| < \varepsilon, \tag{$*$}$$

so when $x \in U$, we have

$$|\varphi(x) - \varphi(x_0)| \leq \int_{K_2} |f(x,y) - f(x_0,y)| d\nu \leq \varepsilon \nu(K_2).$$

Note that $\operatorname{supp} \varphi \subset K_1$, which implies that $\varphi \in C_c(X)$. Then (b) is established. $\qquad \square$

According to Lemma 2.5.3, we define a functional I on $C_c(X \times Y)$: for any $f \in C_c(X \times Y)$,

$$I(f) = \int_X \left(\int_Y f(x,y) d\nu(y) \right) d\mu(x). \tag{2}$$

Obviously, I is a positive linear functional. Then, according to the Riesz representation theorem, there exists a unique Radon measure γ on $X \times Y$ satisfying that for any $f \in C_c(X \times Y)$,

$$I(f) = \int_{X \times Y} f d\gamma. \tag{3}$$

Lemma 2.5.4. *Suppose that X and Y are LCHS, μ and ν are σ-finite Radon measures on X and Y, respectively. If the Radon measure γ is defined by (2) and (3), then γ satisfies (1), that is, for any $A \in \mathscr{B}(X)$ and $B \in \mathscr{B}(Y)$,*

$$\gamma(A \times B) = \mu(A)\nu(B). \tag{4}$$

Proof. We first consider the case that A and B are open sets of X and Y, respectively. For $n \in \mathbb{N}$, take compact sets $K_n \subset A$ and compact sets $J_n \subset B$ such that

$$\lim_{n \to \infty} \mu(K_n) = \mu(A), \quad \lim_{n \to \infty} \nu(J_n) = \nu(B).$$

Take $\varphi_n \prec A$ such that $\chi_{K_n} \leq \varphi_n$ and $\psi_n \prec B$ such that $\chi_{J_n} \leq \psi_n$. Then,

$$\varphi_n \psi_n \prec A \times B.$$

According to the definition of I and γ,

$$\gamma(A \times B) \geq I(\varphi_n \psi_n) = \int_X \varphi_n \left(\int_Y \psi_n d\nu \right) d\mu \geq \mu(K_n)\nu(J_n),$$

$$\mu(A)\nu(B) \geq \gamma(A \times B) \geq \mu(K_n)\nu(J_n).$$

Let $n \to \infty$, we get (4).

Let A, B be the compact sets of X, Y, respectively. Then,

$$\gamma(A \times B) = \inf\{\gamma(G) : G \text{ is an open set and } G \supset A \times B\}.$$

For any open set $G \supset A \times B$ and $x \in X$, define

$$G_x = \{y \in Y : (x, y) \in G\},$$

then G_x is an open subset of Y. When $x \in A$, we have $G_x \supset B$. Take $x \in A$. For any $y \in B$, there exists an open set U_y of X and an open set V_y of Y such that

$$(x, y) \in U_y \times V_y \subset G.$$

The family of open sets $\{V_y : y \in B\}$ is a cover of B. We take a finite subcover $\{V_{y_i} : i = 1, \ldots, m\}$. Denote

$$P_x = \bigcap_{i=1}^{m} U_{y_i}, \quad Q_x = \bigcup_{i=1}^{m} V_{y_i}.$$

Obviously, they are open sets of X, Y, respectively, and

$$\{x\} \times B \subset P_x \times Q_x \subset G.$$

Thus, $\{P_x : x \in A\}$ is an open cover of A. Taking a finite subcover $\{P_{x_i} : i = 1, \ldots, n\}$, write

$$P = \bigcup_{i=1}^{n} P_{x_i}, \quad Q = \bigcap_{i=1}^{n} Q_{x_i}.$$

Then P, Q are open sets of X and Y, respectively, and

$$A \times B \subset P \times Q \subset G.$$

Therefore, we conclude that

$$\gamma(A \times B) = \inf\{\gamma(P \times Q) :$$

P, Q are open sets of X, Y, respectively, and $P \times Q \supset A \times B\}$.

Since $\gamma(P \times Q) = \mu(P)\nu(Q)$ has been proved, when A, B are the compact sets of X and Y, respectively, then (4) holds.

In general case, let $A \in \mathscr{B}(X)$, $B \in \mathscr{B}(Y)$. If $\mu(A)\nu(B) = 0$ or ∞, it is easy to derive (4) by the proven results. Now, let $\mu(A)\nu(B) \in (0, \infty)$. On the one hand, for any $\varepsilon > 0$, there exists a compact set K, an open set U in X, a compact set J and an open set V in Y such that

$$K \subset A \subset U, \ J \subset B \subset V, \ \mu(U \setminus K) + \nu(V \setminus J) < \varepsilon.$$

Then,

$$K \times J \subset A \times B \subset U \times V,$$

$$\gamma(U \times V \setminus K \times J) < \varepsilon(\mu(A) + \nu(B) + \varepsilon).$$

On the other hand,

$$\gamma(K \times J) \leq \gamma(A \times B) \leq \gamma(U \times V),$$

$$\gamma(K \times J) \leq \mu(A)\nu(B) \leq \gamma(U \times V).$$

Since ε is arbitrary, (4) is proved. $\square$

Definition 2.5.1. The Radon measure γ on $X \times Y$ satisfying (4) is called the Radon product of μ and ν, or regular product.

Now, step (i) of answering Question 2 has been completed. Next, we consider step (ii).

For convenience, we prescribe some definitions. Suppose that X and Y are LCHS. If A, B are Borel sets of X and Y, respectively, then $A \times B$ is called a Borel rectangle. If U, V are open sets of X and Y, respectively, then $U \times V$ is called an open rectangle. The collection of all sets that can be expressed as a union of finite number of disjoint Borel rectangles is denoted by $\mathscr{E}$.

Lemma 2.5.5. *Suppose that X and Y are LCHS. If γ is a Radon measure on $X \times Y$, then for each open set $G \subset X \times Y$,*

$$\gamma(G) = \sup\{\gamma(E) : E \in \mathscr{E}, \ E \subset G\}.$$

Proof. Let $0 \leq \alpha < \gamma(G)$. By inner regularity, there exists a compact set $K \subset G$ such that

$$\gamma(K) > \alpha.$$

For each $x \in K$, take an open rectangle R_x such that $x \in R_x \subset G$. Then $\{R_x : x \in K\}$ is an open cover of K. We take a finite subcover from it and denoted it by $\{R_i : i = 1, \ldots, m\}$. Denote

$$E_1 = R_1, \quad E_{i+1} = R_{i+1} \backslash \bigcup \{R_j : j = 1, \ldots, i\}, \quad i = 1, \ldots, m - 1.$$

Obviously, $E_i \in \mathscr{E} \, (i = 1, \ldots, m)$ are pairwise disjoint and

$$\sum_{i=1}^{m} \gamma(E_i) \geq \gamma(K) > \alpha.$$

From this, the required conclusions are derived. $\qquad\square$

Lemma 2.5.6. *Suppose that X and Y are LCHS and μ, ν are Radon measures on X and Y, respectively. Then, there is only one Radon measure satisfying (1) on $X \times Y$.*

Proof. Suppose that γ, τ are Radon measures satisfying (1). Then γ, τ take the same values on each Borel rectangle. Thus, according to Lemma 2.5.5, they take the same values on each open set. By the outer regularity, they are the same. $\qquad\square$

So far, we have completely answered question 2 under the condition that the measure is σ-finite. We describe it as the following theorem.

Theorem 2.5.7. *Suppose that X and Y are LCHS and μ, ν are σ-finite Radon measures on X and Y, respectively. Then, the Radon product (regular product) of μ and ν exists and is unique.*

Exercises 2.1–2.5

1. Let $f_n \in C(X)$, $n \in \mathbb{N}$ and f_n uniformly converges to f, prove that $f \in C(X)$.
2. Let $f, g \in C(X)$, prove that $f + g \in C(X)$, $fg \in C(X)$ and $\min(f, g) \in C(X)$.
3. If $f \in C_c(X)$, prove that f has maximum and minimum values, which also implies that f is bounded.
4. Let X be a Hausdorff space and Y be a subspace of X. Prove that

$$\mathscr{B}(Y) = \{A : A = B \cap Y, \ B \in \mathscr{B}(X)\}.$$

5. Let X be an LCHS. $f \in C(X)$ is said to satisfy the condition $(*)$, which means that for any $\varepsilon > 0$,

$$\{x : |f(x)| \geq \varepsilon\} \text{ is a compact set.} \tag{$*$}$$

Then write

$$C_0(X) = \{f \in C(X) : f \text{ satisfies } (*)\}.$$

Try to prove $C_0(X) = \overline{C_c(X)}$, where $C_0(X)$ is a linear normed space with norm

$$\|f\|_c = \max\{|f(x)| : x \in X\}.$$

6. Let X be an LCHS. Prove that if μ is a σ-finite Radon measure and $E \in \mathscr{B}(X)$, then for any $\varepsilon > 0$, there exist an open set U of X and a closed set F of X such that

$$F \subset E \subset U \text{ and } \mu(U \setminus F) < \varepsilon.$$

7. Let X be an LCHS and K be a compact set of X. Prove that if $I(f)$ is a positive linear functional on $C_c(X)$, then there exists a constant $c(K)$ such that for any $f \in \{f \in C_c(X) : \text{supp}\, f \subset K\}$,

$$|I(f)| \leq c(K)\|f\|_c.$$

8. Prove the conclusion of Lemma 2.4.1 for σ-finite Borel set E can also be established.
9. Prove that Lemma 2.4.1 holds for σ-finite Borel set E, if f is bounded, then g can be taken, satisfying

$$\|g\|_c \leq \sup_{x \in X} |f(x)|.$$

2.6 Haar Measure

2.6.1 *Topological Group*

Let G be a group with respect to multiplication and a topological space. The neighborhood U_x of a point $x \in G$ refers to an open set containing x. The mapping $(x, y) \mapsto xy$ is continuous, that is, if we take any points $x, y \in G$, for any neighborhood U_{xy} of xy, there exists the neighborhood U_x, U_y of x and y such that $uv \in U_{xy}$ with $u \in U_x$, $v \in U_y$. This means that the mapping is continuous if and only if the original of an open set is an open set. The mapping $x \mapsto x^{-1}$ is continuous, which means that for any $x \in G$ and any neighborhood $U_{x^{-1}}$ of x^{-1}, there exists an open set $U_x \ni x$ such that for any $u \in U_x$, $u^{-1} \in U_{x^{-1}}$.

Remark.

$$\begin{cases} \text{The mapping } (x, y) \mapsto xy \text{ is continuous,} \\ \text{The mapping } x \mapsto x^{-1} \text{ is continuous} \end{cases}$$

is equivalent to the mapping $(x, y) \mapsto xy^{-1}$ is continuous.

Proof. In the following, the neighborhood of a point $x, y \in G$ is denoted by U_x and U_y. Suppose that the mapping $(x, y) \mapsto xy^{-1}$ is continuous. Then for the neighborhood $U_{xy^{-1}}$ of xy^{-1}, there exists a neighborhood $U_x \ni x$, $U_y \ni y$ such that $uv^{-1} \in U_{xy^{-1}}$ for any $u \in U_x$, $v \in U_y$. Let $x = u$ be the unit element e of G. It implies that $v^{-1} \in U_{y^{-1}}$ for any $v \in U_y$. This proves that the mapping $y \mapsto y^{-1}$ is continuous. In addition, if we can prove for any $v_1 \in U_{y^{-1}}$, $uv_1 \in U_{xy^{-1}}$, then the mapping $(x, y^{-1}) \mapsto xy^{-1}$ is continuous. In fact, by the continuity of the mapping $y^{-1} \mapsto y$, for the neighborhood U_y of y, there exists a neighborhood $U_{y^{-1}} \ni y^{-1}$ such that for any

$v_1 \in U_{y^{-1}}$, $v_1^{-1} \in U_y$. It can be seen that $uv_1 = u(v_1^{-1})^{-1} \in U_{xy^{-1}}$, which implies that the mapping $(x, y^{-1}) \mapsto xy^{-1}$ is continuous.

Conversely, let

$$\begin{cases} \text{The mapping } (x, y) \mapsto xy \text{ be continuous,} \\ \text{The mapping } x \mapsto x^{-1} \text{ be continuous.} \end{cases}$$

For the neighborhood $U_{xy^{-1}}$ of xy^{-1}, there exists a neighborhood $U_x \ni x$ and $U_{y^{-1}} \ni y^{-1}$ such that for any $u \in U_x$ and $v^{-1} \in U_{y^{-1}}$, $uv^{-1} \in U_{xy^{-1}}$. Note that the mapping $y \mapsto y^{-1}$ is continuous, then for $U_{y^{-1}}$, there exists an open set U_y such that $y \in U_y$ and $v^{-1} \in U_{y^{-1}}$ when $v \in U_y$. Combining all these facts, there exist U_x and U_y such that for any $u \in U_x$ and $v \in U_y$, $uv^{-1} \in U_{xy^{-1}}$, which proves that the mapping $(x, y) \mapsto xy^{-1}$ is continuous.   □

Definition 2.6.1. Let G be a group with respect to multiplication and a topological space. Then G is called a topological group if the mapping $(x, y) \mapsto xy$ from $G \times G$ to G and the mapping $x \mapsto x^{-1}$ from G to G are continuous.

Note that if G is a topological group, then given $u \in G$, $x \mapsto ux$, $x \mapsto xu$ and $x \mapsto x^{-1}$ are all homeomorphic mappings.

We recall some definitions. A topological group G is called a T_0 space if for any $x, y \in G$ with $x \neq y$, there exists an open set U containing only one of x and y. G is a T_1 space if for any two different points $x, y \in G$, there exist two open sets U and V such that $x \in U$, $y \in V$ and $x \notin V$, $y \notin U$, which is equivalent to the single point set of G is closed. We call G is a T_2 space (Hausdorff space) if for any two different points $x, y \in G$, there exist open sets U and V such that $x \in U$, $y \in V$ and $U \cap V = \emptyset$.

Theorem 2.6.1. *Let G be a topological group. Then the following four propositions are equivalent:*

(i) *G is a T_0 space;*
(ii) *G is a T_1 space;*
(iii) *G is a T_2 space;*
(iv) *for the unit element e, $\bigcap \{U : U$ are open sets, $e \in U\} = \{e\}$.*

Proof.   (i)$\Rightarrow$(ii): Let x, y be two distinct points of G. From (i), we know that there exists an open set U which contains only one of x, y.

Let $x \in U$, $y \notin U$ and $V = yU^{-1}x$, then V is an open set and $y \in V$, which implies that $x \notin V$. Otherwise, if there exists $u \in U$ such that $x = yu^{-1}x$, then $y = u \in U$. This is a contradiction.

(ii)$\Rightarrow$(iii): Let $x \neq y$ be two elements of G and $H = G \setminus \{y^{-1}x\}$. Then H is an open set and the unit element $e \in H$. From the continuity of mapping $(u, v) \mapsto uv$ and $ee = e$, we know that there exist neighborhoods U and V of e such that $uv \in H$ when $u \in U$ and $v \in V$. Let

$$W = U \cap U^{-1} \cap V \cap V^{-1}.$$

Then W is an open set and $e \in W = W^{-1}$, $WW^{-1} \subset H$. Suppose that there exist $s, t \in W$ such that $xs = yt$, then

$$x = yts^{-1} \in yWW^{-1} \subset yH = G \setminus \{x\},$$

which causes a contradiction. Therefore, $xW \cap yW = \emptyset$, which proves (iii).

(iii)$\Rightarrow$(iv): For any $x \neq e$, there exist open sets U and V such that $x \in U$, $e \in V$ and $U \cap V = \emptyset$.

(iv)$\Rightarrow$(i): If $x \neq y$, then $x^{-1}y \neq e$. Thus, there exists an open set U such that $e \in U$ and $x^{-1}y \notin U$, that is, $y \notin xU$ and $x \in xU$, then (i) holds. $\qquad\square$

Example 2.6.1. The subspace $\mathbb{R} \setminus \{0\}$ of Euclidean space $\mathbb{R}$ is a locally compact topological group with the usual multiplication and the usual addition, respectively. In addition, the Lebesgue measure m on $\mathbb{R}$ is translation-invariant, that is, for a measurable set $E \subset \mathbb{R}$ and a number $a \in \mathbb{R}$, $m(E) = m(E + a)$.

Example 2.6.2. Let S be a complex number with absolute value (i.e. module) of 1, that is,

$$S = \{e^{i\theta} : \theta \in \mathbb{R}\}.$$

The set S is a compact topological group with the complex multiplication and the relative topology of $\mathbb{R}^2$ (see [11]).

Example 2.6.3. Let M be the whole of the matrix which has the form $\begin{pmatrix} x & y \\ 0 & 1 \end{pmatrix}$, where $x > 0$, $y \in \mathbb{R}$, then M is a group with respect to the usual matrix multiplication. The topology of M is naturally derived from the topology of $\mathbb{R}^2$. Then M is a locally compact topological group.

Example 2.6.4. The rational number set $\mathbb{Q}$ is a topological group with the usual addition and the topology of $\mathbb{R}$. It is a T_2 space but not a locally compact group.

In the following, when talking about a topological group, we assume it is a T_2 space. According to Lemma 2.5.1, this restriction is not strict.

2.6.2   *Continuous Function on the Topological Group*

Let G be a topological group and f be a real-valued (or complex-valued) function on G. The function f is said to be left uniformly continuous if for any $\varepsilon > 0$, there exists a neighborhood U of e such that for $x^{-1}y \in U$,

$$|f(x) - f(y)| < \varepsilon.$$

Similarly, by changing the condition $x^{-1}y \in U$ to $yx^{-1} \in U$, we get the definition of right uniform continuity. Note that for any neighborhood U of e, there exists a symmetric neighborhood $V = V^{-1}$ ($e \in V$) contained in U, then

$$x^{-1}y \in V \Leftrightarrow y^{-1}x \in V,$$

$$yx^{-1} \in V \Leftrightarrow xy^{-1} \in V.$$

It can be seen that the position of x, y in the definition is symmetrical.

Theorem 2.6.2. *Let G be a topological group and $f \in C_c(G)$, then f is both left uniformly continuous and right uniformly continuous, i.e. f is uniformly continuous.*

Proof.   For any $\varepsilon > 0$ and $x \in K := \operatorname{supp} f$, then there exists an open set U_x such that $x \in U_x$ and $|f(x) - f(y)| < \varepsilon$ for $y \in U_x$. Write $W_x = x^{-1}U_x$ and W_x is an open set containing e. So, there exists an open set V_x containing e such that $V_x = V_x^{-1}$ and $V_x \cdot V_x \subset W_x$.

There exist $x_i V_{x_i} (i = 1, 2, \ldots, n)$ such that $\{xV_x : x \in K\} \supset K$. Let

$$V = \bigcap_{k=1}^{n} V_{x_k},$$

then $V = V^{-1}$, $e \in V$. If $x^{-1}y \in V$ and $x \in K$, then there exists $k \in \{1, 2, \ldots, n\}$ such that $x \in x_k V_{x_k}$. Then, on the one hand, $x \in x_k W_{x_k} = U_{x_k}$, we have

$$|f(x) - f(x_k)| < \varepsilon.$$

On the other hand, $y \in xV \subset x_k V_{x_k} \cdot V \subset x_k W_{x_k} = U_{x_k}$, which implies that

$$|f(y) - f(x_k)| < \varepsilon.$$

Thus,

$$|f(x) - f(y)| < 2\varepsilon.$$

If $x^{-1}y \in V$ and $y \in K$, then $|f(x) - f(y)| < 2\varepsilon$ since $y^{-1}x \in V^{-1} = V$ and the results have been proved. If $x^{-1}y \in V$ and $x \notin K$, $y \notin K$, then $f(x) = f(y) = 0$.

In short, it is proved that f is left uniformly continuous. The right uniform continuity of f can be proved using the same argument. $\square$

2.6.3 *The Existence and Uniqueness of Haar Measure*

Definition 2.6.2. Let G be a locally compact group, and μ be a nonzero Radon measure on G. For each set $E \in \mathscr{B}(G)$ and each point $x \in G$, if $\mu(xE) = \mu(E)$, then μ is called a left Haar measure; if $\mu(Ex) = \mu(E)$, then μ is called a right Haar measure. If μ is both a left Haar measure and a right Haar measure, then it is called Haar measure.

The notation xE represents the set $\{u \in G : x^{-1}u \in E\}$ and Ex represents the set $\{u \in G : ux^{-1} \in E\}$.

Theorem 2.6.3. *If G is a locally compact group, then there exists a left Haar measure on G.*

Proof. If K is a compact subset of G, $V \subset G$ and $\overset{\circ}{V} \neq \emptyset$, then

$$K \subset \bigcup_{x \in G} x \overset{\circ}{V}.$$

According to the finite covering property, there exist elements $x_1, \ldots, x_n$ in G such that

$$K \subset \bigcup_{i=1}^{n} x_i \overset{\circ}{V} \subset \bigcup_{i=1}^{n} x_i V.$$

Define

$$(K : V) = \min \left\{ n : \text{there exists } x_1, \ldots, x_n \in G \right.$$

$$\left. \text{such that } K \subset \bigcup_{i=1}^{n} x_i V \right\}.$$

Obviously, $(K : V) = 0$ if and only if $K = \emptyset$.

The set of all compact subsets of G is denoted by $\mathscr{K}$ and the open neighborhood of e is denoted by $\mathscr{U}$. Fix a $K_0 \in \mathscr{K}$ such that $\overset{\circ}{K_0} \neq \emptyset$. For each $U \in \mathscr{U}$, define a real function h_U on $\mathscr{K}$:

$$h_U(K) = (K : U)(K_0 : U)^{-1}.$$

Obviously,

$$\begin{cases} h_U(K_0) = 1, \\ \text{If } K_1 \subset K_2, \text{ then } h_U(K_1) \leq h_U(K_2), \\ h_U(K_1 \cup K_2) \leq h_U(K_1) + h_U(K_2), \\ \text{For any } x \in G, \ h_U(xK) = h_U(K). \end{cases}$$

If $K \subset \bigcup_{i=1}^{m} x_i K_0$ and $K_0 \subset \bigcup_{j=1}^{n} y_j U$ ($U \in \mathscr{U}$), then

$$K \subset \bigcup_{i=1}^{m} \bigcup_{j=1}^{n} (x_i y_j) U,$$

which implies that

$$(K : U) \leq (K : K_0)(K_0 : U).$$

Then

$$0 \leq h_U(K) \leq (K : K_0). \tag{1}$$

We now prove that when $K_1 U^{-1} \cap K_2 U^{-1} = \emptyset$,

$$h_U(K_1 \cup K_2) = h_U(K_1) + h_U(K_2). \tag{2}$$

Let $n = (K_1 \cup K_2 : U)$, $K_1 \cup K_2 \subset \bigcup_{i=1}^n x_i U$ and

$$y_1 \in (x_i U) \cap K_1, \quad y_2 \in (x_i U) \cap K_2,$$

then we have

$$x_i \in (y_1 U^{-1}) \cap (y_2 U^{-1}) \subset (K_1 U^{-1}) \cap (K_2 U^{-1}),$$

which causes a contradiction. Define

$$N_1 = \{i \in \{1, 2, \ldots, n\} : (x_i U) \cap K_1 \neq \emptyset\},$$

$$N_2 = \{i \in \{1, 2, \ldots, n\} : (x_i U) \cap K_2 \neq \emptyset\},$$

then $N_1 \cap N_2 = \emptyset$, which implies that

$$K_1 \subset \bigcup_{i \in N_1} x_i U, \quad K_2 \subset \bigcup_{i \in N_2} x_i U.$$

Therefore,

$$h_U(K_1) + h_U(K_2) \leq h_U(K_1 \cup K_2).$$

Next, we return to the proof of the theorem. For each $K \in \mathscr{K}$, let $I_K = [0, (K : K_0)]$ be a closed interval in $\mathbb{R}$. Define X be the product space $\prod_{K \in \mathscr{K}} I_K$ with product topology. Since each I_K is compact, then X is compact according to the Tihonov theorem.

According to (1), consider the function h_U as a point of X and $h_U(K)$ as the coordinate of point h_U on I_K. For any $V \in \mathscr{U}$, the closure of a subset $\{h_U : U \in \mathscr{U}, U \subset V\}$ of X is denoted by $S(V)$. Obviously, if $V_i \in \mathscr{U}$ and $V = \bigcap_{i=1}^\infty V_i$, then $h_V \in \bigcap_{i=1}^n S(V_i)$, which implies that $S(V) \subset \bigcap_{i=1}^n S(V_i)$. It can be seen that the closed sets $S(U)(U \in \mathscr{U})$ have "finite intersection property". Therefore, $\bigcap_{U \in \mathscr{U}} S(U)$ is not empty. Fixed $h \in S(U)$, then

$$\begin{cases} h(K) \geq 0, \ K \in \mathscr{K}; \ h(\emptyset) = 0, \\ h(K_0) = 1, \\ h(xK) = h(K) \text{ for any } x \in G \text{ and } K \in \mathscr{K}, \\ h(K_1) \leq h(K_2) \text{ if } K_1 \subset K_2, \ K_1, K_2 \in \mathscr{K}, \\ h(K_1 \cup K_2) \leq h(K_1) + h(K_2), \end{cases}$$

where $h(K)$ is used to represent the projection (coordinate) of h on I_K. Claim that if $K_1 \cap K_2 = \emptyset$ $(K_1, K_2 \in \mathscr{K})$,

$$h(K_1 \cup K_2) = h(K_1) + h(K_2). \tag{3}$$

Indeed, since K_1 and K_2 are compact sets, we can find open sets U_i satisfying $U_i \supset K_i (i = 1, 2)$ and $U_1 \cap U_2 = \emptyset$. Then there exists $V \in \mathscr{U}$ such that $K_1 V \subset U_1$ and $K_2 V \subset U_2$. For every $U \in \mathscr{U}$ with $U \subset V^{-1}$, we have

$$(K_1 U^{-1}) \cap (K_2 U^{-1}) = \emptyset.$$

According to (2), it is easy to check

$$h_U(K_1) + h_U(K_2) = h_U(K_1 \cup K_2).$$

Take a point of $q \in X$ whose coordinate on I_K is $q(K)$ $(K \in \mathscr{K})$. We know that the product topology is the minimum topology providing projection continuous. Thus, the mapping $q \mapsto q(K)$ from X into $\mathbb{R}$ is a continuous real function. Fixed $K_1, K_2 \in \mathscr{K}$, define a real function

$$F : q \mapsto q(K_1) + q(K_2) - q(K_1 \cup K_2),$$

then F is continuous. The fact that $F(h_U) = 0$ when $U \in \mathscr{U}$ and $U \subset V^{-1}$ yields that for any $q \in S(V^{-1})$, $F(q) = 0$. Then (3) is proved.

Define the generalized real-valued function μ^* on the topology of G:

$$\mu^*(U) = \sup\{h(K) : K \in \mathscr{K}, \ K \subset U\}.$$

For any $A \subset G$, define

$$\mu^*(A) = \inf \{\mu^*(U) : A \subset U, \ U \text{ is an open set}\}.$$

Now, we verify that μ^* is an outer measure on G.

On the one hand, $\mu^*(\emptyset) = 0$ and μ^* is monotonically increasing.

On the other hand, what we need to do is to verify the countable subadditivity of the open set family and then consider the general situation. Let U_i be an open set of G,

$$U = \bigcup_{i=1}^{\infty} U_i.$$

If K is a compact subset of U, then there exists $n \in \mathbb{N}$ such that $K \subset \bigcup_{i=1}^{n} U_i$. According to Theorems 2.2.1–2.2.3, we know that there exists $f_i \in C_c(G)$ satisfying for any $x \in K$,

$$0 \le f_i \le 1; \ \operatorname{supp} f_i \subset U_i, \ \sum_{i=1}^{n} f_i(x) = 1.$$

Let $K_i = K \cap \operatorname{supp} f_i$, then

$$K_i \in \mathscr{K}, \ K_i \subset U_i; \quad K = \bigcup_{i=1}^{n} K_i,$$

which implies that

$$h(K) \le \sum_{i=1}^{n} h(K_i) \le \sum_{i=1}^{n} \mu^*(U_i) \le \sum_{i=1}^{\infty} \mu^*(U_i).$$

Therefore,

$$\mu^*(U) \le \sum_{i=1}^{\infty} \mu^*(U_i).$$

Take $A_i \subset G$ and let $A = \sum_{i=1}^{\infty} A_i$. For any given $\varepsilon > 0$, there exist open sets U_i such that $A_i \subset U_i$ and

$$\mu^*(U_i) \le \mu^*(A_i) + 2^{-i}\varepsilon.$$

Then

$$\mu^*(A) \le \mu^* \left(\bigcup_{i=1}^{\infty} U_i \right) \le \sum_{i=1}^{\infty} \mu^*(U_i) \le \sum_{i=1}^{\infty} \mu^*(A_i) + \varepsilon,$$

which proves the countable subadditivity of μ^*. Thus, μ^* is an outer measure.

Now, we verify that any open set U is μ^*-measurable, i.e. U satisfies the Carathéodory condition with respect to μ^*. Let V be an open set and $\mu^*(V) < \infty$, then for any given $\varepsilon > 0$, we can choose a compact set K such that $K \subset V \cap U$ satisfying

$$h(K) > \mu^*(V \cap U) - \varepsilon.$$

Then, we select a compact set $L \subset V \cap \complement K$ such that

$$h(L) > \mu^*(V \cap \complement K) - \varepsilon.$$

Using the fact that $K \cap L = \emptyset$, we obtain

$$h(K \cup L) = h(K) + h(L) > \mu^*(V \cap U) + \mu^*(V \cap \complement K) - 2\varepsilon$$
$$\ge \mu^*(V \cap U)) + \mu^*(V \cap \complement U) - 2\varepsilon.$$

Thus,

$$\mu^*(V) \ge \mu^*(V \cap U) + \mu^*(V \cap \complement U).$$

For $\varepsilon > 0$ and any set $A \subset G$ with $\mu^*(A) < \infty$, then there exists an open set $V \supset A$ such that

$$\mu^*(A) > \mu^*(V) - \varepsilon.$$

Then, combining all these facts, we have

$$\mu^*(A) > \mu^*(V \cap U) + \mu^*(V \cap \complement U) - \varepsilon$$
$$\ge \mu^*(A \cap U) + \mu^*(A \cap \complement U) - \varepsilon.$$

Therefore,

$$\mu^*(A) \ge \mu^*(A \cap U) + \mu^*(A \cap \complement U).$$

Now, we have proved that the open set U is μ^*-measurable, which indicates that every element of $\mathscr{B}(G)$ (Borel algebra on G) is μ^*-measurable. The restriction of μ^* on $\mathscr{B}(G)$ is denoted by μ, which is a measure.

If K is a compact set, then according to Theorem 2.2.1, there exists an open set U such that $K = U$ and $\overline{U}$ is a compact set. Then

$$\mu(K) \leq \mu(U) = \sup\{h(A) : A \in \mathscr{K}, \ A \subset U\} \leq h(\overline{U}).$$

Obviously, $h(\overline{U}) < \infty$, which yields $\mu(K) < \infty$, that is, μ takes finite values on a compact set. It can be seen that μ makes every measurable set have outer regularity. If U is an open set containing a compact set K, then $h(K) \leq \mu(U)$. From this, we deduce that

$$h(K) \leq \inf\{\mu(U) : U \text{ is an open set and } U \supset K\} = \mu(K).$$

Thus, for any open set U, the following equalities,

$$\mu(U) = \sup\{h(K) : K \in \mathscr{K}, \ K \subset U\}$$
$$= \sup\{\mu(K) : K \in \mathscr{K}, \ K \subset U\},$$

hold, which proves that every open set is inner regular.

It is easy to check that $\mu(K_0) \geq h(K_0) = 1$. It can be seen that μ is not constant zero, and the left invariance of μ is obvious.

We have proved that μ is a left Haar measure. $\qquad\square$

Theorem 2.6.4. *Let G be a locally compact group and μ be the left Haar measure on G, then*

(1) *if U contained in G is a nonempty open set, then $\mu(U) > 0$;*
(2) *if $f \in C_c(G)$, $f \geq 0$, $f \neq 0$ (i.e. $f \not\equiv 0$), then*

$$\int_G f \, d\mu > 0.$$

Proof. Let K be a compact set with $\mu(K) > 0$. For a nonempty open set $U \subset G$, the open set family $\{xU : x \in G\}$ is an open cover of K, then there exist $x_1, \ldots, x_n \in G$ such that $(\bigcup_{i=1}^n x_i U) \supset K$. Therefore,

$$0 < \mu(K) \leq \sum_{i=1}^n \mu(x_i U) = n\mu(U).$$

Then (1) is proved.

Let $f \in C_c(G)$ satisfy the conditions of (2). Then there exists a nonempty open set U and a positive number ε such that $f \geq \varepsilon \chi_U$. This implies that

$$\int_G f \, d\mu \geq \varepsilon \mu(U) > 0.$$

$\square$

Theorem 2.6.5. *Let G be a locally compact group and μ, ν be two left Haar measures on G. Then there exists a positive number a such that $\mu = a\nu$.*

Proof.   Fix $g \in C_c(G)$ such that $g \geq 0$ and $g \not\equiv 0$. For any $f \in C_c(G)$, let

$$h(x, y) = \frac{f(x)g(yx)}{\int_G g(tx) \, d\nu(t)}, \quad (x, y) \in G \times G,$$

then h is a continuous function with compact support on $G \times G$. It is easy to verify that the interchanges of integrals are well justified by the Fubini theorem (see Section 2.5):

$$\int \left[\int h(x, y) \, d\nu(y) \right] d\mu(x)$$

$$= \int f(x) \frac{\int g(yx) \, d\nu(y)}{\int g(tx) \, d\nu(t)} d\mu(x)$$

$$= \int f(x) \, d\mu(x) = \int \left[\int h(x, y) \, d\mu(x) \right] d\nu(y)$$

and

$$\int \left[\int \frac{f(x)g(yx)}{\int g(tx) \, d\nu(t)} d\mu(x) \right] d\nu(y)$$

$$= \int \left[\int \frac{f(y^{-1}x)g(x)}{\int g(ty^{-1}x) \, d\nu(t)} d\mu(x) \right] d\nu(y).$$

Next, by interchanges of integrals,

$$\int \left[\int \frac{f(y^{-1}x)g(x)}{\int g(ty^{-1}x) \, d\nu(t)} d\nu(y) \right] d\mu(x)$$

$$= \int \left[\int \frac{f(y^{-1})g(x)}{\int g(ty^{-1}) \, d\nu(t)} d\nu(y) \right] d\mu(x)$$

$$= \int g(x)\, d\mu(x) \cdot \int \frac{f(y^{-1})}{\int g(ty^{-1})\, d\nu(t)}\, d\nu(y).$$

Write

$$\lambda(f,g;\nu) = \int \frac{f(y^{-1})}{\int g(ty^{-1})\, d\nu(t)}\, d\nu(y).$$

We have

$$\int f\, d\mu = \int g\, d\mu \cdot \lambda(f,g;\nu).$$

This fact implies that

$$\int f\, d\nu = \int g\, d\nu \cdot \lambda(f,g;\nu),$$

where we replace μ by ν.

Let $a = \dfrac{\int g\, d\mu}{\int g\, d\nu}$, then we have

$$\int f\, d\mu = a \cdot \int g\, d\nu \cdot \lambda(f,g;\nu) = a \int f\, d\nu.$$

Note that $a > 0$ is independent of f. Since $f \in C_c(G)$ is arbitrary, we can conclude that $\mu = a\nu$. $\qquad\square$

2.6.4 The Properties of Haar Measure

We emphasize again that when talking about topological groups, the topology of this group is Hausdorff topology.

Let μ be a measure on a group and A be a measurable set and we denote $\check{\mu}(A) = \mu(A^{-1})$, then $\check{\mu}$ is also a Radon measure.

Theorem 2.6.6. *Let G be a locally compact group and μ be a Radon measure on G. Then μ is a left Haar measure $\Leftrightarrow$ $\check{\mu}$ is a right Haar measure.*

Proof. The proof is trivial. $\qquad\square$

It can be seen from Theorems 2.6.5 and 2.6.6 that the right Haar measures are multiples of each other. In fact, since our discussion of the left Haar measure does not depend on the left multiplication property of the group G, we only need to change "left" to "right" in the discussion of the right Haar measure.

Theorem 2.6.7. *Let G be a locally compact group and μ be a left Haar measure on G. Then μ is a finite measure $\Leftrightarrow G$ is compact.*

Proof.   Let $\mu(G) < \infty$. Take a compact set K such that $\mu(K) > 0$. Let $\{xK : x \in G\}$ be a family of compact sets. Inductively, finite number of $x_i K$, $i = 1, \ldots, n$, can be extracted with the property that they are pairwise disjoint and any other xK must intersect with $\bigcup_{i=1}^n (x_i K)$, otherwise it will contradict $\mu(G) < \infty$. Then

$$G = \bigcup_{i=1}^n (x_i K) \cdot K^{-1}$$

is compact.   □

Let G be a locally compact group and μ be a left Haar measure on G. Due to the fact that the mapping $u \mapsto ux$ (fix $x \in G$) is a self-homeomorphism of G, it defines a Radon measure on G by $\mu_x(A) = \mu(Ax)$. Since μ is left translation-invariant, μ_x also has this property, and then μ_x is also a left Haar measure, which implies that $\mu_x = \Delta(x)\mu$, where $\Delta(x)$ is a positive number related to x. Let ν be an another left Haar measure on G, then there exists a positive number a such that $\nu = a\mu$. Thus,

$$\nu_x = a\mu_x = a\Delta(x)\mu = \Delta(x)\nu.$$

It can be seen that $\Delta(x)$ is independent of μ and is uniquely determined by the group G.

Definition 2.6.3. The above positive function $\Delta(x)$ on G is called the right module function of G.

Let f be a function on G and f_x denote the right translation x $(x \in G)$ of f, i.e.

$$f_x(t) = f(tx^{-1}), \quad t \in G.$$

Similarly, $_x f$ denotes the left translation x of f, that is, $_x f(t) = f(x^{-1}t)$, $t \in G$. If A is a Borel set of a locally compact group G, μ

is a left Haar measure on G, and $\Delta(x)$ is a right module function of G, then we have

$$\int_G (\chi_A)_x \, d\mu = \mu(Ax) = \Delta(x)\mu(A) = \Delta(x) \int_G \chi_A \, d\mu$$

since $(\chi_A)_x = \chi_{Ax}$. For a nonnegative Borel measurable function or a μ integrable function f, the following equality holds:

$$\int f_x \, d\mu = \Delta(x) \int f \, d\mu \left(\text{or} \int f(t) \, d\mu(t) = \Delta(x) \int f(tx) \, d\mu(t) \right).$$

Theorem 2.6.8. *Let G be a locally compact group and Δ be a right module function of G. Then*

(1) *Δ is continuous;*
(2) *for any $x, y \in G$, $\Delta(xy) = \Delta(x)\Delta(y)$.*

Proof. Take a left Haar measure μ and a nonnegative function $f \in C_c(G)$ such that $f \not\equiv 0$. Write $a = \int f \, d\mu > 0$ and $K = \operatorname{supp} f$. Fix $x_0 \in G$. We now prove that Δ is continuous at x_0.

Given any $\varepsilon > 0$, by the uniform continuity of f (see Theorem 2.6.2), there exists a neighborhood V_ε of identity element e such that when $s \in tV$,

$$|f(s) - f(t)| < \varepsilon.$$

$\overline{V_\varepsilon}$ can be considered to be compact and $V_\varepsilon = V_\varepsilon^{-1}$. So, when $x \in V_\varepsilon x_0$, we have

$$tx^{-1} \in (tx_0^{-1})V_\varepsilon.$$

Then for any $t \in G$,

$$|f(tx^{-1}) - f(tx_0^{-1})| < \varepsilon.$$

When $t \notin K\overline{V_\varepsilon}x_0$, $tx_0^{-1} \notin K\overline{V_\varepsilon}$ and for $x \in V_\varepsilon x_0$, we have $x^{-1} \in x_0^{-1}V_\varepsilon$. Thus, $tx^{-1} \notin K$. By the above relationships, $f(tx^{-1}) = f(tx_0^{-1}) = 0$. This implies that

$$\left| \int f(tx^{-1}) \, d\mu(t) - \int f(tx_0^{-1}) \, d\mu(t) \right|$$

$$\leq \int |f(tx^{-1}) - f(tx_0^{-1})| \, d\mu(t) \leq \varepsilon \cdot \mu(K\overline{V_\varepsilon}x_0).$$

For all $\varepsilon \in (0,1)$, assume that $V_\varepsilon \subset V_1$, then $\mu(K\overline{V_1}x_0)$ is a constant independent of $\varepsilon \in (0,1)$, this yields that the continuity of

$$\int f(tx^{-1})\, d\mu(t) = \Delta(x)a$$

at x_0. Thus, the conclusion (1) is proved.

For any Borel set A, we have

$$\Delta(xy)\mu(A) = \mu(Axy) = \Delta(y)\mu(Ax) = \Delta(y)\Delta(x)\mu(A),$$

which yields the conclusion (2).   □

Obviously, we can define a left module function $\delta(x)$ in a symmetric way, and it is easy to verify that $\delta(x) = [\Delta(x)]^{-1} = \Delta(x^{-1})$. Therefore, it is enough to study only the right module.

Definition 2.6.4. If a right module function $\Delta(x)$ on a locally compact group G is equal to 1, then we call G is unimodular.

Obviously, a locally compact group G is unimodular $\Leftrightarrow$ its every left Haar measure is a right Haar measure.

The locally compact group which is interchangeable is unimodular.

Theorem 2.6.9. *A compact group is unimodular.*

Proof. Let $\Delta(x)$ be a right module function of a compact group G. Then, for any $x \in G$ and $n \in \mathbb{N}$, $\Delta(x^n) = [\Delta(x)]^n$. If there exists $x \in G$ such that $\Delta(x) \neq 1$, then one of $\Delta(x)$ and $\Delta(x^{-1})$ is greater than 1. Furthermore, Δ is a continuous and unbounded function on the compact set G, which is impossible.   □

The following assertion is obvious.

Theorem 2.6.10. *Let G be a locally compact group and μ be a left Haar measure on G. Then for each $A \in \mathscr{B}(G)$,*

$$\check{\mu}(A) = \int_A \Delta(x^{-1})\, d\mu(x).$$

Exercise 2.6

1. Let $\mathbb{Q}$ be the additive group of the rational numbers, which inherits the topology of $\mathbb{R}$ (see [11]). Prove that there does not exist nonzero translation-invariant Radon measure on $\mathbb{Q}$.

2. Assume G is the multiplicative group of the positive real numbers, which inherits the topology of $\mathbb{R}$. Prove that $\mu(A) = \int_A \frac{1}{x}\,dx$ defines a Haar measure on G.

3. Suppose G is a locally compact group and μ is a left Haar measure on G. Prove that the topology of G is discrete $\Leftrightarrow$ for some $x \in G$, $\mu(\{x\}) > 0$.

4. Let G is a locally compact group, μ is a right Haar measure on G and Δ be a right module function. Prove that for each measurable set A and $x \in G$, we have $\mu(xA) = \Delta(x^{-1})\mu(A)$.

5. Assume G is a locally compact group and μ is a left Haar measure on G. Prove that G is unimodular $\Leftrightarrow \mu = \check{\mu}$.

6. Suppose G is a locally compact group and μ is a left Haar measure on G. Prove that μ is σ finite $\Leftrightarrow G$ is σ compact.

7. Let G is a locally compact group, μ is a left Haar measure on G and ν be a right Haar measure on G. Prove that if a measurable set A satisfies $\mu(A) = 0$, then $\nu(A) = 0$.

Chapter 3

Lebesgue Integration on $\mathbb{R}^n$

The rest of this book is the deepening and development of theory of functions of real variable. The focus is to introduce the modern theories and methods of real analysis on $\mathbb{R}^n$. For this purpose, we emphasize the important theories of operator interpolation and maximal function. This role has been throughout the contents of the following chapters. In detail, we regard the theory of operator interpolation as a powerful tool for establishing various integral transform norm inequalities and regard the maximal function as an effective method to deal with convergence problems (a.e. and nontangential convergence). These theories and methods reflect the most essential things in the development of modern analysis.

On the other hand, the Lebesgue theory on $\mathbb{R}^n$ provides not only the most important example for the abstract integral theory in the first chapter but also one of the most important examples with countable bases for the theory of locally compact Hausdorff spaces in the second chapter.

In the remainder of this book, we use m to denote the Lebesgue measure on $\mathbb{R}^n$. The integral element $dm(x)$ under Lebesgue integral is abbreviated as dx. When the measurable set or measurability is mentioned, in the absence of ambiguity, it always refers to Lebesgue measurable set or Lebesgue measurability.

For a measurable set E, notations mE, $m(E)$, and $|E|$ are all used to denote its measure. For $x = (x_1, \ldots, x_n)$ in $\mathbb{R}^n$, the notation $|x|$ denotes its Euclidean norm, that is, $|x| = (x_1^2 + \cdots + x_n^2)^{\frac{1}{2}}$.

3.1   Integral Calculation under Linear Transformation

When we talk about the substitution of Riemann integral

$$\int_{-\infty}^{\infty} f(x)dx = \int_{-\infty}^{\infty} f(\psi(t))\psi'(t)dt$$

on the real line $\mathbb{R}$, it is necessary to involve the properties of the function (transformation) $\psi : \mathbb{R} \to \mathbb{R}$. When it comes to the variable substitution of Lebesgue integral on $\mathbb{R}^n$, transformation T must be able to map a Lebesgue measurable set in $\mathbb{R}^n$ to another Lebesgue measurable set in $\mathbb{R}^n$.

Definition 3.1.1. Let A be a measurable set in $\mathbb{R}^n$ and T be a mapping from A to $\mathbb{R}^n$. We refer to such maps as transformations on A. T is called a measurable transformation if the image of every measurable set under T is measurable.

Note that T is not necessarily measurable even if it is continuous. For example, let C be a Cantor set on $\mathbb{R}$:

$$C = \left\{ x \in [0,1] : x = \sum_{j=1}^{\infty} a_j 3^{-j}, a_j = 0, 2 \text{ for any } j \in \mathbb{N} \right\}.$$

Define a transformation $T : C \longrightarrow [0,1]$ as

$$T(x) = \sum_{j=1}^{\infty} a_j 2^{-j-1} \text{ when } x = \sum_{j=1}^{\infty} a_j 3^{-j} \in C.$$

Obviously, T is a continuous bijection from C to $[0,1]$. Taking a nonmeasurable set $E \subseteq [0,1]$ and $C_0 = T^{-1}(E) \subset C$, we get $m(C_0) = 0$, which means that T maps the measurable set C_0 in $\mathbb{R}$ to the nonmeasurable set E in $\mathbb{R}$. However, the occurrence of the above phenomenon can be avoided if we slightly strengthen the conditions of T.

Definition 3.1.2. Let T be a transformation on $A \subset \mathbb{R}^n$. T is called a Lipschitz transformation on A, which is also denoted by Lip-transformation if

$$|Tx - Ty| \le c|x - y| \text{ for any } x, y \in A, \tag{L}$$

where $c > 0$ is independent of x and y.

Remark. A transformation is continuous if the preimage of every open set is an open set. It is easy to verify that condition (L) implies continuity. Thus, a Lip-transformation must be continuous.

Theorem 3.1.1. *A Lip-transformation on a measurable set is a measurable transformation.*

Proof. Let T be a Lip-transformation on a measurable set. First, by the continuity, T maps a compact set to a compact set (the proof is left to readers). Furthermore, T maps an F_σ set to an F_σ set because an F_σ set is a countable union of compact sets.

Next, we note that T maps a null set to a null set. To see this, let I be a cube in $\mathbb{R}^n$. Then by the definition of Lip-transformation, there exists $c > 0$ such that

$$m^*(TI) \le c|I|,$$

where m^* is the Lebesgue outer measure on $\mathbb{R}^n$. If $m(Z) = 0$, then there exists a countable series of closed cubes $\{I_k\}$ such that $\bigcup_k I_k \supset Z$ and $\sum_k |I_k| < \varepsilon$. Moreover, we have $\bigcup_k TI_k \supset TZ$ and

$$m^*(TZ) \le \sum_k m^*(TI_k) \le c \sum_k |I_k| < c\varepsilon.$$

Hence, $m(TZ) = 0$.

Finally, for any measurable set E in the domain of T, E can be represented as $E = H \bigcup Z$, where H is an F_σ set and Z is a null set. Moreover, TH is an F_σ set and FZ is a null set, hence $TE = TH \bigcup TZ$ is a measurable set. $\qquad\square$

In this section, we only focus on linear transformations, one of the simplest class of Lip-transformations. For a linear transformation T, there is a simple quantitative relationship between $m(E)$ and $m(TE)$.

Theorem 3.1.2. *Let $T : \mathbb{R}^n \to \mathbb{R}^n$ be an invertible linear transformation and E be a measurable set. Then we have*

$$m(TE) = |\det T|m(E).$$

Remark. For a measurable set E, if we define

$$\mu(E) = m(TE),$$

then obviously μ is a measure. Theorem 3.1.2 means that $\mu \ll m$, μ is σ-finite and the Radon–Nikodym derivative

$$\frac{d\mu}{dm} = |\det T|.$$

We now give the proof of Theorem 3.1.2.

Proof. Using the fact that a linear transformation T could be denoted by a matrix of order $n \times n$, let $T = (a_{ij})_{n\times n}$, $x = (x_1, x_2, \ldots, x_n) \in \mathbb{R}^n$, then

$$Tx = \left(\sum_{i=1}^{n} a_{i1}x_i, \sum_{i=1}^{n} a_{i2}x_i, \ldots, \sum_{i=1}^{n} a_{in}x_i \right).$$

Now, we consider three special linear transformations (elementary transformation):

(a) T_1 multiplies only one component of x by a nonzero constant α, while leaving the other components unchanged.
(b) T_2 adds a nonzero constant multiple of another component to only one component of x while leaving the other components unchanged.
(c) T_3 swaps the positions of two components of x while leaving the other components unchanged.

These three linear transformations have following simple properties:

$$\det T_1 = \alpha, \quad \det T_2 = 1 \quad \text{and} \quad \det T_3 = -1.$$

It is well known that if T is an invertible linear transformation, then T could be split into $T = T^{(1)}T^{(2)} \cdots T^{(m)}$, where $T^{(k)}, 1 \leq k \leq m$ must be one of T_1, T_2 and T_3. Thus, it remains to be proved that the theorem holds for elementary transformations.

Suppose that Q is a cube with each side parallel to coordinate axis and T is an elementary transformation. It is easy to verify that

$$m(TQ) = |\det T| m(Q).$$

According to the definition of measure, for a measurable set E,

$$m(TE) \le \inf \left\{ \sum_{k=1}^{\infty} m(TQ_k) : Q_k \text{ is a cube and } \bigcup_{k=1}^{\infty} Q_k \supset E \right\}$$

$$= |\det T| m(E).$$

Evidently, T^{-1} and T belong to the same class of elementary transformation. Using the above equation on the transformation T^{-1} and the measurable set TE, we obtain the reverse inequality

$$m(T^{-1}TE) = m(E) \le |\det T^{-1}| m(TE),$$

thereby finishing the proof. $\square$

Next, we consider the substitution of integral.

Lemma 3.1.3. *Suppose that A and B are measurable sets of $\mathbb{R}^n$ and T is an invertible transformation from A to B with T and T^{-1} measurable. Define a measure ν on A by*

$$\nu(E) = m(TE),$$

where $E \subset A$ is measurable. Then for all nonnegative measurable function f on B,

$$\int_A f(Tx)d\nu(x) = \int_B f(x)dx.$$

Proof. We only need to consider characteristic function $f = \chi_E$, $E \subset B$ which follows immediately by

$$\int_B \chi_E(x)dx = m(E) = \nu(T^{-1}E) = \int_A \chi_E(Tx)d\nu(x).$$

$\square$

Now, we consider the integral calculation under linear transformation.

Theorem 3.1.4. *Let $T : \mathbb{R}^n \to \mathbb{R}^n$ be an invertible linear transformation. If $f \in L(\mathbb{R}^n)$, then*

$$\int_{\mathbb{R}^n} f(x)dx = |\det T| \int_{\mathbb{R}^n} f(Tx)dx.$$

Proof.   Define ν as Lemma 3.1.3, then

$$\int_{\mathbb{R}^n} f(Tx)d\nu(x) = \int_{\mathbb{R}^n} f(x)dx.$$

But from Theorem 3.1.2, ν is σ-finite, $\nu \ll m$ and

$$\frac{d\nu}{dm} = |\det T|.$$

Hence, by the Radon–Nikodym theorem (see Section 1.5), we have

$$\int_{\mathbb{R}^n} f(x)dx = \int_{\mathbb{R}^n} f(Tx)d\nu(x) = |\det T| \int_{\mathbb{R}^n} f(Tx)dx.$$

$\square$

Corollary 3.1.5. *The Lebesgue measure m is invariable under orthogonal transformation ρ, i.e. $m(\rho E) = m(E)$.*

Proof.   Let ρ be an orthogonal transformation, i.e. a linear transformation ρ satisfying $\rho\rho^* = I$, where ρ^* is the transposition of ρ and I is identity transformation. Thus, $|\det \rho| = 1$, and the conclusion follows by Theorem 3.1.2. $\square$

Corollary 3.1.6. *Let $\delta_a : x \mapsto ax, a > 0$ be a dilation operator, then*

$$m(\delta_a E) = a^n m(E).$$

Proof.   It follows immediately by $|\det \delta_a| = a^n$ and Theorem 3.1.2.

$\square$

3.2   Integral Calculation under Regular Transformation

In this section, we extend the substitution of integral under linear transformation in Section 3.1 to a more general case under continuous transformation.

Definition 3.2.1. Let V be an open set in $\mathbb{R}^n$. Transformation $T : V \to \mathbb{R}^n$ is defined by

$$T(x) = \big(t_1(x), t_2(x), \ldots, t_n(x)\big), x \in V,$$

where $t_j(x), 1 \leq j \leq n$ are functions defined on V. Transformation T is called a C^1 transformation on V if every $t_j(x)$ has the first-order continuous partial derivative $\frac{\partial t_j}{\partial x_i}(x), 1 \leq i, j \leq n$, i.e. $t_j \in C^1(V)$, $j = 1, \ldots, n$. The matrix consisting of $n \times n$ partial derivatives is denoted by $D(T)(x)$, i.e.

$$D(T)(x) = \left(\frac{\partial t_j(x)}{\partial x_i}\right)_{1 \leq i, j \leq n}, \quad x \in V.$$

As we did in multivariable differential calculus, the Jacobi determinant identity of T is denoted by $J_T(x) = \det D(T)(x)$.

Definition 3.2.2. Let T be a C^1 transformation on $V \subset \mathbb{R}^n$. We call T a regular transformation on V if T is injective and $J_T(x) \neq 0$ for all $x \in V$.

Remark. Knowing from multivariable differential calculus [15], if T is a regular transformation on V, then $T^{-1} : T(V) \to V$ will also be a regular transformation on $T(V)$ and for any $x \in T(V)$,

$$D(T^{-1})(x) \cdot D(T)(T^{-1}x) = I,$$

where I denotes the identity matrix.

First, we want to know if a regular transformation T is measurable.

Lemma 3.2.1. *All* C^1 *transformations are measurable.*

Proof. By the topological properties of $\mathbb{R}^n$ (it is an LCHS), every open set V can be rewritten as

$$V = \bigcup_{k=1}^{\infty} B_k, \text{ where } B_k \text{ are open balls with finite radius and } \overline{B_k} \subset V.$$

Let

$$M_k = \max\left\{\left|\frac{\partial t_j(x)}{\partial x_i}\right| : x \in \overline{B_k}, i, j = 1, \ldots, n\right\}.$$

Obviously, $M_k < \infty$. By mean value theorem, for any $x, x' \in \overline{B_k}$

$$|t_j(x) - t_j(x')| \leq nM_k|x - x'|, \quad j = 1, \ldots, n.$$

Thus,

$$|T(x) - T(x')| \leq n^2 M_k|x - x'|,$$

which indicates that T is a Lip-transformation on B_k. Thereby, T maps measurable subsets of B_k to measurable sets, which holds for all $k \in \mathbb{N}$. Using the above decomposition of V, we conclude that T is a measurable transformation on V. □

Definition 3.2.3. Let $M(x) = \big(m_{ij}(x)\big)_{n \times n}$ be a matrix on $E \subset \mathbb{R}^n$. We define

$$|M(x)| = \max\left\{\sum_{j=1}^{n} |m_{ij}(x)| : i = 1, \ldots, n\right\},$$

$$\|M\|_E = \sup\{|M(x)| : x \in E\}.$$

Lemma 3.2.2. *Suppose that* $x^0 = (x_1^0, \ldots, x_n^0) \in \mathbb{R}^n, 0 < h < \infty$ *and*

$$Q = \{(x_1, \ldots, x_n) : x_j^0 - h \le x_j < x_j^0 + h, j = 1, \ldots, n\}.$$

If T *is a* C^1 *transformation on an open set* V *and* $\overline{Q} \subset V$, *then* TQ *is contained in a closed cube of side length* $2h\|D(T)\|_Q$ *centered at* Tx^0.

Proof. Let $Tx = (t_1(x), \ldots, t_n(x)), x \in V$. By mean value theorem, for any $x \in Q$ and $i \in \{1, \ldots, n\}$, there exists y^i on the line segment joining x and x^0 such that

$$t_i(x) - t_i(x^0) = \sum_{j=1}^{n} \frac{\partial t_i}{\partial x_j}(y^i)(x_j - x_j^0).$$

Hence,

$$|t_i(x) - t_i(x^0)| \le h \sup\left\{\sum_{j=1}^{n} \left|\frac{\partial t_i}{\partial y_j}(y)\right| : y \in Q\right\},$$

which draws the conclusion. □

Theorem 3.2.3. *Let* T *be a regular transformation on an open set* V. *If we define* ν *as Lemma 3.1.3, i.e. for all measurable sets* $E \subset V$,

$$\nu(E) = m(TE),$$

then ν *is a* σ-*finite measure,* $\nu \ll m$ *and the Radon–Nikodym derivative*

$$\frac{d\nu}{dm}(x) = |J_T(x)|, \ x \in V.$$

Proof. We will first prove that for all cubes Q as described in Lemma 3.2.2, if $\overline{Q} \subset V$, then

$$\nu(Q) \leq \int_Q |J_T(x)|dx. \tag{3.1}$$

Take such a cube Q. Note that the related functions are uniformly continuous on the compact set $\overline{Q}$, i.e. for all $\varepsilon \in (0, \frac{1}{4})$, there exists $\delta > 0$ such that

$$|(D(T)(y))^{-1}(D(T)(x) - D(T)(y))| < \varepsilon, \tag{3.2}$$

whenever $x, y \in Q$ and $|x - y| < \delta$. Take $m \in \mathbb{N}$ big enough such that $\frac{2h}{m} < \frac{\delta}{n}$. Now, divide Q into $N = m^n$ congruent subcubes. Each small cube is represented by

$$\left\{ (x_1, \ldots, x_n) : x_j^0 - h + \frac{2(k_j - 1)h}{m} \right.$$
$$\left. \leq x_j < x_j^0 - h + \frac{2k_j h}{m}, j = 1, \ldots, n \right\},$$

where $(k_1, \ldots, k_n) \in \{1, \ldots, m\}^n$. These small cubes are denoted by $Q_k, k = 1, \ldots, N$. The center of Q_k is denoted by x_k.

The basic idea of proof is to approximate T on Q_k with the compound of a linear transformation L_k and an appropriate translation, where the linear transformation L_k on Q_k is defined by

$$L_k = D(T)(x^k), \quad k = 1, \ldots, N.$$

Condition $J_T(x) \neq 0$ makes sure that L_k^{-1} exists. Obviously, $L_k^{-1}T$ is still the C^1 transformation on V. We now prove that the compound of $L_k^{-1}T$ and an appropriate translation is indeed approximate to identity transformation I on Q_k (when m is very large). Define

$$\Delta_k = L_k^{-1}T - I = L_k^{-1}(T - L_k), \quad a^k = \Delta_k(x^k), \quad k = 1, \ldots, N,$$

and transformation δ_k by

$$\delta_k(x) = \Delta_k(x) - a^k \quad \text{for all} \quad x \in V, \quad k = 1, \ldots, N.$$

We can see that δ_k is still the C^1 transformation and for all $x \in V$,

$$L_k^{-1}T(x) = I(x) + \delta_k(x) + a^k. \tag{3.3}$$

The laws of differentiation and matrix multiplication give that

$$D(\delta_k) = D(\Delta_k) = L_k^{-1}(D(T) - L_k).$$

Note that when $x \in Q_k$,

$$|x - x^k| < \frac{2nh}{m} < \delta.$$

Using the definition of L_k and (3.2), we may obtain

$$\|\delta_k\|_{Q_k} \le \varepsilon, \quad k = 1, \ldots, N.$$

Note that $\delta_k(x^k) = O$, where O is the origin. By Lemma 3.2.2, $\delta_k(Q_k)$ is contained in a cube of side length $\frac{2h}{m}\varepsilon$ centered at the origin. Thus, by (3.3), $L^{-1}TQ_k$ must be contained in a cube of side length $(1 + 2\varepsilon)\frac{2h}{m}$ centered at a^k. Furthermore,

$$m(L_k^{-1}TQ_k) \le (1 + 2\varepsilon)^n m(Q_k), \quad k = 1, \ldots, N. \tag{3.4}$$

By Theorem 3.1.2, for $k = 1, \ldots, N$,

$$\nu(Q_k) = m(L_k L_k^{-1} T Q_k) = |\det L_k| m(L_k^{-1} T Q_k). \tag{3.5}$$

Combining (3.4) and (3.5) with $\det L_k = J_T(x^k)$,

$$\nu(Q) \le (1 + 2\varepsilon)^n \sum_{k=1}^N |J_T(x^k)| m(Q_k). \tag{3.6}$$

It is easy to verify that the sum in the equation above is just a Riemann sum of integral $\int_Q |J_T(x)| dx$. Let $N \to \infty$ in (3.6),

$$\nu(Q) \le (1 + 2\varepsilon)^n \int_Q |J_T(x)| dx,$$

then (3.1) follows by letting $\varepsilon \to 0$.

Define a measure μ,

$$\mu(E) = \int_E |J_T(x)| dx \text{ for any measurable set } E \subset V.$$

By (3.1), we get that for any measurable set $E \subset V$,

$$\nu(E) \le \mu(E), \tag{3.7}$$

which concludes that ν is a σ-finite measure and $\nu \ll m$. By (3.7) and the Radon–Nikodym theorem, for any nonnegative measurable function f on TV,

$$\int_V f(Tx)d\nu(x) \le \int_V f(Tx)d\mu(x) = \int_V f(Tx)|J_T(x)|dx.$$

By Lemma 3.1.3,

$$\int_V f(Tx)d\nu(x) = \int_{TV} f(x)dx.$$

Therefore,

$$\int_{TV} f(x)dx \le \int_V f(Tx)|J_T(x)|dx. \tag{3.8}$$

Since T is a regular transformation, we may use (3.8) for T^{-1} to get

$$\int_{TV} f(x)|J_T(T^{-1}x)||J_{T^{-1}}(x)|dx \ge \int_V f(Tx)|J_T(x)|dx.$$

By the Remark of Definition 3.2.2, $|J_T(T^{-1}x)||J_{T^{-1}}(x)| = 1$ for all $x \in TV$. Thus, (8) is actually an equality. Moreover, (3.7) is an equality. □

As an immediate result of Theorem 3.2.3 and Lemma 3.1.3, we have the following result about the substitution of integral.

Theorem 3.2.4. *Let T be a regular transformation on V. If f is a nonnegative measurable function on TV or $f \in L(TV)$, then*

$$\int_{TV} f(x)dx = \int_V f(Tx)|J_T(x)|dx.$$

3.3 Integral Calculation under Spherical Coordinates

Let $n \ge 2$ be an integer. Define

$$\Sigma_{n-1} = \{x \in \mathbb{R}^n : |x| = (x_1^2 + \cdots + x_n^2)^{\frac{1}{2}} = 1\}.$$

Suppose that m is a Lebesgue measure on $\mathbb{R}^n$. We have claimed that when referring to Lebesgue measure on $\mathbb{R}^n$ in this section, it refers to the usual Lebesgue measure in the theory of functions of a real variable, which has been discussed in detail. If we denote all

Lebesgue measurable sets in $\mathbb{R}^n$ by $\mathscr{M}$, then $(\mathbb{R}^n, \mathscr{M}, m)$ is a σ-finite measure space. Let $\mathscr{B}(\mathbb{R}^n)$ be the family of all Borel sets in $\mathbb{R}^n$ and the restriction of m on $\mathscr{B}(\mathbb{R}^n)$ is a Radon measure.

Let $X = \mathbb{R}^n \backslash \{O\}$ (O is the origin) and $Y = (0, \infty) \times \Sigma_{n-1}$. For every $x \in X$, there exists a unique point $(|x|, \frac{x}{|x|}) \in Y$. This correspondence is denoted by φ, i.e. $\varphi(x) = (|x|, \frac{x}{|x|})$. It is easy to verify that φ is invertible: $\varphi^{-1}(r, \xi) = r\xi, 0 < r < \infty, \xi \in \Sigma_{n-1}$. Define that the open sets of Y are the images of all open sets of X under φ, then φ is a homeomorphism from X to Y. $\varphi(x)$ is called the spherical coordinate of x ($x \neq 0$). X is identical to Y in the sense of up to the invertible map φ.

Definition 3.3.1. Let $E \subset \Sigma_{n-1}$. We say that E is a (Lebesgue) measurable set of X if $\varphi^{-1}((0,1) \times E)$ is a measurable set of X (a measurable set of $\mathbb{R}^n$). Let $\mathscr{E}$ be all measurable sets on Σ_{n-1}. Consider a function σ on $\mathscr{E}$: for any $E \in \mathscr{E}$,

$$\sigma(E) = nm(\varphi^{-1}((0,1) \times E)).$$

We refer to $\sigma(E)$ as the (Lebesgue) measure of E.

Remark. We can only define $\sigma(G) = nm(\varphi^{-1}(0,1] \times G)$ for open set G on Σ_{n-1} (intersection of an open set of $\mathbb{R}^n$ and Σ_{n-1}), which induces an outer measure σ^* on Σ_{n-1}:

$$\sigma^*(E) = \inf\{\sigma(G) : G \text{ is open and } G \supset E\}.$$

Then use Carathéodory's condition to get the measurable sets on Σ_{n-1} and the related measure, which can be proved to be the same as Definition 3.3.1.

Let $\aleph$ be all measurable sets in $(0, \infty)$ (intersection of measurable sets in $\mathbb{R}$ and $(0, \infty)$). Define a measure ρ on $\aleph$ to be

$$\rho(A) = \int_A r^{n-1} dr \text{ for any } A \in \aleph,$$

where dr is the usual Lebesgue measure on $\mathbb{R}$. Obviously, ρ is absolutely continuous with respect to Lebesgue measure on $(0, \infty)$. ρ is σ-finite and complete.

Hence, as we mentioned in Chapter 1, there exists a product measure $(Y, \mathscr{M}', \rho \times \sigma)$ induced by measure spaces $((0, \infty), \aleph, \rho)$ and $(\Sigma_{n-1}, \mathscr{E}, \sigma)$.

Theorem 3.3.1. *In the sense of homeomorphism* φ,

$$\mathcal{M}' = \mathcal{M}, \quad \rho \times \sigma = m.$$

Proof. Let $\mathcal{R} = \{A \times B : A \in \mathcal{N}, B \in \mathcal{E}\}$ be the family of all measurable rectangles on Y. By the definition of $\rho \times \sigma$,

$$\rho \times \sigma(A \times B) = \rho(A)\sigma(B), \quad A \times B \in \mathcal{R}.$$

If

$$m(\varphi^{-1}(A \times B)) = \rho(A)\sigma(B), \quad A \times B \in \mathcal{R} \qquad (3.9)$$

can be proved, then we may conclude that $\rho \times \sigma$ agrees with m on $\mathcal{R}$. Furthermore, it can be deduced by the geometric structure of $\mathbb{R}^n$ that the outer measure $(\rho \times \sigma)^* = m^*$, hence $\mathcal{M}' = \mathcal{M}, \rho \times \sigma = m$. Thus, it suffices to prove (3.9).

Let $B \in \mathcal{E}$. We consider the case of $A = (a, b)(0 \le a < b < \infty)$. At this moment,

$$A \times B = (0, b) \times B \backslash (0, a] \times B.$$

Thus,

$$m(\varphi^{-1}(A \times B)) = m(\varphi^{-1}((0, b) \times B)) - m(\varphi^{-1}((0, a] \times B)).$$

Denote $E(a) = \varphi^{-1}((0, a) \times B)(a > 0)$. Let δ_a be a dilation operator on $\mathbb{R}^n$ (see Section 3.1), then by Corollary 3.1.6 and the definition of σ,

$$m(E(a)) = m(\delta_a E(1)) = a^n m(E(1)) = \frac{1}{n} a^n \sigma(B).$$

Thus,

$$m(\varphi^{-1}(A \times B)) = \frac{1}{n}(b^n - a^n)\sigma(B) = \rho((a, b))\sigma(B).$$

It can be seen from this that m coincides with $\rho \times \sigma$ on $\aleph \times B$.

Since B is arbitrary, we may deduce that m coincides with $\rho \times \sigma$ on $\mathcal{R}$. $\qquad \square$

The result of Theorem 3.3.1 may be rewritten as

$$(\mathbb{R}^n, \mathcal{M}, m) = ((0, \infty) \times \Sigma_{n-1}, \mathcal{M}, \rho \times \sigma)$$

or just $m = \rho \times \sigma$ in the sense of homeomorphism φ.

Corollary 3.3.2. *Let* $f \in L(\mathbb{R}^n)$, *then*

$$\int_{\mathbb{R}^n} f(x)dx = \int_0^\infty \int_{\Sigma_{n-1}} f(r\xi)d\sigma(\xi)r^{n-1}dr, \qquad (3.10)$$

which is the integral calculation formula under spherical coordinates.

Proof. We only need to consider characteristic functions. Let E be a measurable set in $\mathbb{R}^n$ and still denote $E\backslash\{O\} \subset X$ by E. Using Theorem 3.3.1, we obtain

$$\int_{\mathbb{R}^n} \chi_E dm = \int_X \chi_E dm = m(E) = \rho \times \sigma(\varphi(E))$$

$$= \int_Y \chi_{\varphi(E)}(y)d(\rho \times \sigma)(y)$$

$$= \int_Y \chi_E(\varphi^{-1}(y))d(\rho \times \sigma)(y).$$

For $y = (r, \xi)$, we have $\varphi^{-1}(y) = r\xi$. Using equation

$$\frac{d(\rho \times \sigma)}{dm}(r, \xi) = r^{n-1},$$

we finish the proof. $\qquad\qquad\qquad\qquad\qquad\qquad\qquad\qquad\square$

Next corollary follows immediately from Corollary 3.3.2.

Corollary 3.3.3. *Suppose that* $f \in L(\mathbb{R}^n)$ *and* $f(x) = g(|x|)$, *then* $g \in L((0, \infty), \aleph, \rho)$ *and*

$$\int_{\mathbb{R}^n} f(x)dx = \sigma(\Sigma_{n-1}) \int_0^\infty g(r)r^{n-1}dr.$$

We say that f *is a radial function if* $f(x) = g(|x|)$.

3.4   Extension of Two Important Inequalities

In Section 1.3, two important integral inequalities, Hölder's inequality and Minkowski's inequality, have been established. We first consider generalized Hölder's inequality.

Theorem 3.4.1. *Suppose that*

$$\frac{1}{p_1} + \frac{1}{p_2} + \cdots + \frac{1}{p_m} = 1, \quad 1 \le \frac{1}{p_i} \le \infty \quad (1 \le i \le m).$$

Let $f_i \in L^{p_i}(\mathbb{R}^n)(1 \le i \le m)$, *then*

$$\int_{\mathbb{R}^n} |f_1(x) \cdots f_m(x)| dx \le \|f_1\|_{p_1} \cdots \|f_m\|_{p_m}.$$

Proof. It is Hölder's inequality when $m = 2$. The method of proof of cases $m > 3$ is the same as the case $m = 3$, so we just need to consider the case $m = 3$. Note that

$$\left(\frac{p_1 p_2}{p_1 + p_2}\right)^{-1} + p_3^{-1} = 1.$$

Using Hölder's inequality,

$$\int_{\mathbb{R}^n} |f_1(x) f_2(x) f_3(x)| dx \le \left(\int_{\mathbb{R}^n} |f_1(x) f_2(x)|^{\frac{p_1 p_2}{p_1 + p_2}} dx\right)^{\frac{p_1 + p_2}{p_1 p_2}} \|f_3\|_{p_3}.$$

Note that

$$\frac{p_2}{p_1 + p_2} + \frac{p_1}{p_1 + p_2} = 1.$$

Using Hölder's inequality again for the integral inside the parentheses of the right-hand side of the above inequality, we get

$$\int_{\mathbb{R}^n} |f_1(x) f_2(x)|^{\frac{p_1 p_2}{p_1 + p_2}} dx \le \|f_1\|_{p_1}^{\frac{p_1 p_2}{p_1 + p_2}} \cdot \|f_2\|_{p_2}^{\frac{p_1 p_2}{p_1 + p_2}}.$$

The conclusion follows immediately from the above two inequalities. $\square$

Suppose that $1 \le p < \infty$, $\frac{1}{p} + \frac{1}{p'} = 1$, $f \in L^p(\mathbb{R}^n)$ and $g \in L^{p'}(\mathbb{R}^n)$, then Hölder's inequality

$$\left|\int_{\mathbb{R}^n} f(x) g(x) dx\right| \le \|f\|_p \|g\|_{p'}$$

indicates that $\|f\|_p$ is an upper bound of the set $\{|\int_{\mathbb{R}^n} f(x) g(x) dx| : \|g\|_{p'} = 1\}$. However, if take

$$g = \operatorname{sgn} f |f|^{p-1} \|f\|_p^{1-p} \text{ when } \|f\|_p \ne 0,$$

then we can see

$$\|g\|_{p'} = 1, \quad \left|\int_{\mathbb{R}^n} f(x) g(x) dx\right| = \|f\|_p,$$

which yields that

$$\|f\|_p = \sup \left\{ \left| \int_{\mathbb{R}^n} f(x)g(x)dx \right| : \|g\|_{p'} = 1 \right\}.$$

The following result is a more precise form of the above conclusion and it is more convenient in use.

Theorem 3.4.2. *Let* $1 \le p < \infty, \frac{1}{p} + \frac{1}{p'} = 1$. *If* $f \in L^p(\mathbb{R}^n)$, *then*

$$\|f\|_p = \sup \left\{ \left| \int_{\mathbb{R}^n} f(x)g(x)dx \right| : \|g\|_{p'} = 1, g \in D \right\},$$

where D *is the set of simple functions.*

Proof.   Since $|f(x)|$ is nonnegative and measurable, there exists a sequence of nonnegative functions $\{\varphi_m\} \subset D$ such that $\varphi_m \nearrow |f|$. If we denote

$$f_m = \varphi_m \operatorname{sgn} f,$$

then $f_m \in D$, $|f_m| \le |f|$ and

$$\lim_{m \to \infty} f_m(x) = f(x), \text{ a.e.}$$

Setting

$$g_m = |f_m|^{p-1} \|f_m\|_p^{1-p} \operatorname{sgn} f_m$$

yields that $g_m \in D$, $\|g_m\|_{p'} = 1$ and

$$\left| \int_{\mathbb{R}^n} f_m(x)g_m(x)dx \right| = \|f_m\|_p.$$

Fatou's lemma implies that

$$\|f\|_p = \|\lim_{m \to \infty} f_m\|_p \le \varliminf_{m \to \infty} \|f_m\|_p$$

$$= \lim_{m \to \infty} \left| \int_{\mathbb{R}^n} f_m(x)g_m(x)dx \right|$$

$$\le \lim_{m \to \infty} \int_{\mathbb{R}^n} |f_m(x)g_m(x)|dx$$

$$= \lim_{m \to \infty} \int_{\mathbb{R}^n} |f(x)g_m(x)|dx$$

$$= \lim_{m \to \infty} \left| \int_{\mathbb{R}^n} f(x)g_m(x)dx \right|.$$

Thus,

$$\|f\|_p \leq \sup\left\{\left|\int_{\mathbb{R}^n} f(x)g(x)dx\right| : \|g\|_{p'} = 1, g \in D\right\}.$$

The reversed inequality is obvious. $\square$

The following theorem is an extension of Minkowski's inequality.

Theorem 3.4.3 (Minkowski's inequality for integrals). *Suppose that* $1 \leq p < \infty$, $f(x, y)$ *is Lebesgue measurable on* $\mathbb{R}^{m+n} = \mathbb{R}^m \times \mathbb{R}^n, m, n \in \mathbb{N}$, *then*

$$\left\{\int_{\mathbb{R}^m} \left[\int_{\mathbb{R}^n} |f(x, y)|dy\right]^p dx\right\}^{\frac{1}{p}} \leq \int_{\mathbb{R}^n} \left(\int_{\mathbb{R}^m} |f(x, y)|^p dx\right)^{\frac{1}{p}} dy.$$

Proof. When $p = 1$, the conclusion follows from Fubini's theorem. Let

$$1 < p < \infty, \quad \frac{1}{p} + \frac{1}{p'} = 1,$$

$0 \leq g \in L^{p'}(\mathbb{R}^m)$ and $\|g\|_{p'} = 1$. The Tonelli theorem and Hölder's inequality tell us that

$$\int_{\mathbb{R}^m} \left(\int_{\mathbb{R}^n} |f(x, y)|dy\right) g(x)dx = \int_{\mathbb{R}^n} \left(\int_{\mathbb{R}^m} |f(x, y)|g(x)dx\right) dy$$

$$\leq \int_{\mathbb{R}^n} \left(\int_{\mathbb{R}^m} |f(x, y)|^p dx\right)^{\frac{1}{p}} dy \cdot \|g\|_{p'}.$$

Thus,

$$\sup_{\|g\|_{p'}=1} \int_{\mathbb{R}^m} \left(\int_{\mathbb{R}^n} |f(x, y)|dy\right) g(x)dx \leq \int_{\mathbb{R}^n} \left(\int_{\mathbb{R}^m} |f(x, y)|^p dx\right)^{\frac{1}{p}} dy,$$

which is the inequality we want to prove. $\square$

Exercise 3

1. Take $Z \subset \mathbb{R}$ with $m(Z) = 0$. Denote $W = \{x^2 : x \in Z\}$. Prove that $m(W) = 0$.

2. Let T be a Lip-transformation and a C^1-transformation on $\mathbb{R}^n$. Prove that

$$\sup_{x \in \mathbb{R}^n} |J_T(x)| < \infty.$$

3. Prove that the map $x \mapsto D(T)(x)$ is a continuous map on V if T is a C^1-transformation on V.

4. When T is an invertible linear transformation, prove that T is a regular transformation, and for any $x \in \mathbb{R}^n$,

$$D(T)(x) = T.$$

5. Prove that

$$\sigma(\Sigma_{n-1}) = \frac{2\pi^{\frac{n}{2}}}{\Gamma(\frac{n}{2})}.$$

6. If $a > 0$, then

$$\int_{\mathbb{R}^n} \exp(-a|x|^2)dx = \left(\frac{\pi}{a}\right)^{\frac{n}{2}}.$$

7. Define

$$P(x) = c_n(1 + |x|^2)^{\frac{-(n+1)}{2}},$$

where

$$c_n = \pi^{\frac{-(n+1)}{2}} \Gamma\left(\frac{n+1}{2}\right).$$

Prove that

$$\int_{\mathbb{R}^n} P(x)dx = 1.$$

Chapter 4

The Interpolation of Operator on $L^p(\mathbb{R}^n)$

When studying functions, it is common to transform the examined object f into another function $T(f)$ based on a specific corresponding rule T. The properties of f are reflected by the properties of $T(f)$. We call the corresponding rule (or mapping) T as an operator and the specific definition can be found in [3, 8, 13]. For example, the maximal function, convolution, and Fourier transform discussed in Chapters 5–7 involve three important operators, respectively. In the study of operators, the problem of boundedness is the most basic problem. The concept of boundedness is as follows.

We denote by $\mathscr{F}$ the set of measurable functions on $\mathbb{R}^n$. L^p stands for $L^p(\mathbb{R}^n)$, and its norm (when $1 \le p \le \infty$) is denoted by $\|\cdot\|_p$.

Definition 4.0.1. The operator $T : L^p(\mathbb{R}^n) \to L^q(\mathbb{R}^n), 1 \le p, q \le \infty$, is called type (p, q) (or (p, q) bounded) if it satisfies

$$\|T(f)\|_q \le C\|f\|_p,$$

where the constant C is independent of f. The smallest constant C satisfying the above inequality is called the (p, q) norm of T and denoted by $\|T\|_{(p,q)}$.

Definition 4.0.2. For $1 \le p, q \le \infty$, the operator $T : L^p(\mathbb{R}^n) \to \mathscr{F}$ is called weak type (p, q) (or weak (p, q) bounded) if

(i) for $1 \le q < \infty$, there exists a constant $C > 0$ such that

$$\sup_{\alpha>0} \alpha \left[m(\{x \in \mathbb{R}^n : |T(f)(x)| > \alpha\}) \right]^{\frac{1}{q}} \le C\|f\|_p,$$

(ii) for $q = \infty$, the following inequality holds;

$$\|T(f)\|_q \leq C\|f\|_p,$$

where the constant C is independent of f and $m(E)$ is the Lebesgue measure of the set $E \subset \mathbb{R}^n$. The smallest constant C satisfying the above inequality is called weak (p, q) norm of T and denoted by $\|T\|_{w(p,q)}$.

Obviously, when $q = \infty$, the operator T being of weak type (p, q) is equivalent to T being of type (p, q). Besides, by inequality

$$\int_{\{x \in \mathbb{R}^n : |T(f)(x)| > \alpha\}} |T(f)(x)|^q dx \leq \|T(f)\|_q^q,$$

the operator of type (p, q) must be of weak type (p, q) and $\|T\|_{w(p,q)} \leq \|T\|_{(p,q)}$. Since the operator T which we deal with must have some properties related to linear operation, it is always required that the domain X of T is a linear subset of $\mathscr{F}$. It should be pointed out that if $1 \leq p < r < q \leq \infty$ and $X \supset L^p, X \supset L^q$, then since every $f \in L^r$ must be able to be represented as

$$f = f_1 + f_2, \quad f_1 \in L^p, \quad f_2 \in L^q,$$

X as a linear set must contain L^r.

Can the boundedness of T on L^p and L^q determine boundedness on L^r? This is called the interpolation of operators.

The interpolation of linear operators will be discussed in Section 4.1 and the interpolation of sublinear operators will be studied in Section 4.2.

4.1   The Riesz–Thörin Theorem

Let X be a linear subset of $\mathscr{F}$, the operator $T : X \to \mathscr{F}$ is called linear if for any $a, b \in \mathbb{R}$, and for any $f, g \in X$,

$$T(af + bg) = aT(f) + bT(g).$$

Theorem 4.1.1 (Riesz–Thörin's theorem). *Let $1 \le p_j, q_j \le \infty$, $j = 0, 1$, $t \in (0, 1)$ and*

$$\frac{1}{p_t} = \frac{1-t}{p_0} + \frac{t}{p_1}, \quad \frac{1}{q_t} = \frac{1-t}{q_0} + \frac{t}{q_1}.$$

If a linear operator T is of type (p_j, q_j), $j = 0, 1$, then T is also of type (p_t, q_t) and satisfies

$$\|T\|_{(p_t, q_t)} \le \|T\|_{(p_0, q_0)}^{1-t} \|T\|_{(p_1, q_1)}^{t}.$$

In order to prove Theorem 4.1.1, the following two key lemmas are given first.

Lemma 4.1.2 (Phragmen–Lindelöf three lines theorem). *Suppose*

$$S = \{z = x + iy \in \mathbb{C} : 0 \le x \le 1, y \in \mathbb{R}\}$$

is a closed subset in the complex plane and the complex-valued function F on S is continuous and bounded on S as well as analytic in $\mathring{S}$. If on the boundary of S, F satisfies

$$|F(iy)| \le K_0, \ |F(1 + iy)| \le K_1, \ y \in \mathbb{R},$$

then for any $x \in (0, 1)$ and $y \in \mathbb{R}$, the following inequality holds in $\mathring{S}$:

$$|F(x + iy)| \le K_0^{1-x} K_1^x, \ x \in (0, 1), \ y \in \mathbb{R}.$$

Proof. Assume that $K_0, K_1 > 0$ and define $G(z)$ by

$$G(z) = \frac{F(z)}{K_0^{1-z} K_1^z},$$

then the problem remains to prove the following proposition: If G is continuous and bounded on S, as well as analytic in $\mathring{S}$, and for any

$y \in \mathbb{R}$,

$$|G(iy)| \leq 1, \quad |G(1+iy)| \leq 1,$$

then, for any $x \in (0,1)$ and $y \in \mathbb{R}$, we have

$$|G(x+iy)| \leq 1.$$

Firstly, consider the special case:

$$\lim_{|y|\to\infty} \max \left\{ |G(x+iy)| : x \in [0,1] \right\} = 0.$$

Obviously, in this case, there exists a $y_0 > 0$ such that for $x \in [0,1]$ and $|y| \geq y_0$,

$$|G(x+iy)| \leq 1. \qquad (*)$$

Thus, on the boundary ∂Q of the rectangle Q with iy_0, $1+iy_0, 1-iy_0, -iy_0$ as vertices, we have

$$|G(z)| \leq 1, \quad z \in \partial Q.$$

The maximum modulus principle of analytic function implies that

$$|G(z)| \leq 1, \quad z \in Q.$$

It follows from $(*)$ that

$$|G(z)| \leq 1, \quad z \in S\backslash Q.$$

Therefore, for the above special case, the conclusion holds.

Secondly, for the general case, apply the above results to the function

$$G_m(z) = G(z)e^{\frac{z^2-1}{m}} \quad (m \in \mathbb{N}),$$

which implies that for any $z \in S$,

$$|G_m(z)| \leq 1.$$

Then, let $m \to \infty$, we get the conclusion

$$|G(z)| \leq 1, \quad z \in S.$$

$\square$

Lemma 4.1.3. *If the conclusion of Theorem 4.1.1 holds for simple functions, then Theorem 4.1.1 holds.*

Proof.   Without loss of generality, we assume that $p_0 \leq p_1$, let

$$p = p_t, q = q_t, \; k_0 = \|T\|_{(p_0,q_0)},$$
$$k_1 = \|T\|_{(p_1,q_1)}, \; M = k_0^{1-t} k_1^t.$$

For $f \in L^p$, define

$$E = \{x \in \mathbb{R}^n : |f(x)| > 1\},$$
$$F = \{x \in \mathbb{R}^n : |f(x)| \leq 1\},$$
$$f^0 = f\chi_E, \; f^1 = f\chi_F.$$

Obviously,

$$f = f^0 + f^1, \; f^0 \in L^{p_0} \cap L^p, \; f^1 \in L^{p_1} \cap L^p.$$

Denote D the set of all integrable simple functions and take $g_m \in D$, $m \in \mathbb{N}$ such that

$$\lim_{m \to \infty} \|g_m - f^0\|_p = 0; \; g_m = g_m\chi_E,$$

and at the same time, take $h_m \in D, m \in \mathbb{N}$ such that

$$\lim_{m \to \infty} \|h_m - f^1\|_p = 0; \; h_m = h_m\chi_F, \; h_m f^1 \geq 0.$$

Since

$$\|g_m - f^0\|_{p_0} \leq |E|^{\frac{1}{p_0} - \frac{1}{p}} \|g_m - f^0\|_p, \; \|h_m - f^1\|_{p_1} \leq (\|h_m - f^1\|_p)^{\frac{p}{p_1}},$$

and the linear operator T is of types (p_0, q_0) and (p_1, q_1), we can deduce that

$$\lim_{m \to \infty} (\|T(g_m - f^0)\|_{q_0} + \|T(h_m - f^1)\|_{q_1}) = 0.$$

Thus, $\{g_m\}$ contains a subsequence, still written as $\{g_m\}$ satisfying

$$\lim_{m \to \infty} |T(g_m)(x) - T(f^0)(x)| = 0, \; a.e. \; x \in \mathbb{R}^n.$$

Similarly, $\{h_m\}$ contains a subsequence, still written as $\{h_m\}$ such that

$$\lim_{m\to\infty} |T(h_m)(x) - T(f^1)(x)| = 0, \ \ a.e. \ x \in \mathbb{R}^n.$$

Denote $f_m = g_m + h_m$, by the linear property of T, we get

$$\lim_{m\to\infty} |T(f_m)(x) - T(f)(x)| = 0, \ \ a.e. \ x \in \mathbb{R}^n.$$

Obviously, the following equality holds:

$$\lim_{m\to\infty} \|f_m - f\|_p = \lim_{m\to\infty} (\|g_m - f^0\|_p^p + \|h_m - f^1\|_p^p)^{\frac{1}{p}} = 0,$$

therefore, we have

$$\lim_{m\to\infty} \|f_m\|_p = \|f\|_p.$$

According to the assumption, we obtain

$$\|T(f_m)\|_q \le M\|f_m\|_p, \ m \in \mathbb{N}.$$

Then, by Fatou's theorem (see Theorem 1.2.8),

$$\|T(f)\|_q \le \varliminf_{m\to\infty} \|T(f_m)\|_q \le \varliminf_{m\to\infty} M\|f_m\|_p = M\|f\|_p.$$

$\square$

Proof of Theorem 4.1.1. We only prove the situations $(p_0, p_1) \ne (\infty, \infty)$ and $(q_0, q_1) \ne (1, 1)$. The proof for other cases is similar and left to the reader.

Denote the associated number of p by p', that is, when $p = 1$, $p' = \infty$; when $p = \infty$, $p' = 1$; and when $1 < p < \infty$, $\frac{1}{p} + \frac{1}{p'} = 1$. Now, we write

$$\alpha_j = \frac{1}{p_j}, \ \beta_j = \frac{1}{q_j}, j = 0, 1;$$

$$\alpha = \frac{1}{p_t}, \ \beta = \frac{1}{q_t};$$

$$\alpha(z) = (1 - z)\alpha_0 + z\alpha_1, \ \beta(z) = (1 - z)\beta_0 + z\beta_1.$$

It is easy to know that

$$\alpha(j) = \alpha_j, \ \beta(j) = \beta_j, \ j = 0, 1;$$

$$\alpha(t) = \alpha, \ \beta(t) = \beta.$$

For simplicity, write

$$\|T\|_{(p_i, q_i)} = k_i, \ i = 0, 1; \ M = k_0^{1-t} k_1^t; \ p_t = p, \ q_t = q.$$

Using the notations above, the inequality to be proved in the theorem becomes, for any $f \in L^p$,

$$\|T(f)\|_q \leq M\|f\|_p. \tag{4.1}$$

Still denote the set of all integrable simple functions by D. According to Lemma 4.1.3, the necessary and sufficient condition for (4.1) is that for any $f \in D$,

$$\|T(f)\|_q \leq M\|f\|_p. \tag{4.1'}$$

From Theorem 3.4.2, we have

$$\|T(f)\|_q = \sup_{g \in D, \|g\|_{q'} = 1} \left| \int_{\mathbb{R}^n} T(f)(u)g(u)du \right|.$$

Thus, (4.1') holds if and only if for any $f, g \in D$ with $\|f\|_p = \|g\|_{q'} = 1$,

$$\left| \int_{\mathbb{R}^n} T(f)(u)g(u)du \right| \leq M. \tag{4.2}$$

Define

$$f_z(u) = e^{i \arg f} |f(u)|^{\frac{\alpha(z)}{\alpha}},$$

$$g_z(u) = e^{i \arg g} |g(u)|^{\frac{1-\beta(z)}{1-\beta}},$$

$$F(z) = \int_{R^n} T(f_z)(u) g_z(u) du.$$

Note that

$$f_t(u) = f(u), \quad g_t(u) = g(u),$$

thus (4.2) is equivalent to for any $f, g \in D$ with $\|f\|_p = \|g\|_{q'} = 1$,

$$|F(t)| \leq M. \tag{4.3}$$

It is easy to see that $F(z)$ is a bounded analytic function on the closed region S. From Lemma 4.1.2, in order to prove (4.3), it suffices to verify

$$|F(iy)| \leq k_0, \quad |F(1+iy)| \leq k_1.$$

Here, only the proof of the first inequality is given (the second is similar). In fact, by $\alpha(iy) = \alpha_0 + iy(\alpha_1 - \alpha_0)$ and $1 - \beta(iy) = (1 - \beta_0) - iy(\beta_1 - \beta_0)$, we can know that

$$|f_{iy}(u)|^{p_0} = \left| e^{i \arg f} |f(u)|^{\frac{\alpha(iy)}{\alpha}} \right|^{p_0} = \left| |f(u)|^{iy \frac{\alpha_1 - \alpha_0}{\alpha}} |f(u)|^{\frac{p}{p_0}} \right|^{p_0} = |f(u)|^p$$

and

$$|g_{iy}(u)|^{q_0'} = |g(u)|^{q'}.$$

Thus, the conditions of the theorem and Hölder's inequality give

$$|F(iy)| \leq \|T f_{iy}\|_{q_0} \|g_{iy}\|_{q_0'} \leq k_0 \|f_{iy}\|_{p_0} \|g_{iy}\|_{q_0'} = k_0 \|f\|_p^{\frac{p}{p_0}} \|g\|_{q'}^{\frac{q'}{q_0'}} = k_0,$$

which confirms (4.3), so (4.1) is valid. $\square$

4.2 The Marcinkiewicz Theorem

Definition 4.2.1 (Quasi-linear operator). An operator T, defined on a linear subset X of $\mathscr{F}$ is called quasi-linear if there exists a constant $A > 0$ such that for any $f_1, f_2 \in X$ and $x \in \mathbb{R}^n$,

$$|T(f_1 + f_2)(x)| \leq A\big(|T(f_1)(x)| + |T(f_2)(x)|\big),$$

and when $A = 1$, T is called sub-linear.

Obviously, a linear operator must be sub-linear.

Theorem 4.2.1 (Marcinkiewicz's theorem). *Let* $1 \leq p_j \leq q_j \leq \infty$ *$(j = 0, 1)$, $q_0 \neq q_1$ and for $t \in (0, 1)$,*

$$\frac{1}{p_t} = \frac{1-t}{p_0} + \frac{t}{p_1}; \quad \frac{1}{q_t} = \frac{1-t}{q_0} + \frac{t}{q_1}.$$

Assume that T is a sub-linear operator. If T is of weak type (p_j, q_j), $(j = 0, 1)$, then T is of type (p_t, q_t) and

$$\|T\|_{(p_t, q_t)} \leq C\|T\|_{w(p_0, q_0)}^{1-t}\|T\|_{w(p_1, q_1)}^{t},$$

where the constant C depends on p_0, q_0, p_1, q_1 and t.

Obviously, the conditions of Theorem 4.2.1 are weaker than Theorem 4.1.1. However, the conclusions of these two theorems are generally incomparable. Therefore, Theorem 4.2.1 is not a generalization of Theorem 4.1.1.

For simplicity, only the cases $p_0 = q_0 = 1$ and $p_1 = q_1 = \infty$ are given here. For the general case, the proof method is the same. The result in this special case is written as Theorem 4.2.2.

Theorem 4.2.2. *Let a sub-linear operator T be of weak type $(1, 1)$ and type (∞, ∞), then T is of type (p, p), $1 < p < \infty$ and*

$$\|T\|_{(p,p)} \leq \frac{2p}{p-1}\|T\|_{w(1,1)}^{\frac{1}{p}}\|T\|_{(\infty,\infty)}^{1-\frac{1}{p}}.$$

Proof. Assume that $f \in L^p(\mathbb{R}^n)$, $1 < p < \infty$. Take $\alpha > 0$ and define

$$f_1(x) = \begin{cases} f(x), & |f(x)| \le \alpha, \\ \alpha \operatorname{sgn} f(x), & |f(x)| > \alpha, \end{cases}$$

and

$$f_2(x) = \begin{cases} 0, & |f(x)| \le \alpha, \\ (|f(x)| - \alpha) \cdot \operatorname{sgn} f(x), & |f(x)| > \alpha, \end{cases}$$

then $f(x) = f_1(x) + f_2(x)$. On the one hand, by $\|f_1\|_\infty \le \alpha$ and

$$\|f_2\|_1 = \int_{\{x \in \mathbb{R}^n : |f(x)| > \alpha\}} (|f(x)| - \alpha) dx$$

$$\le \int_{\{x \in \mathbb{R}^n : |f(x)| > \alpha\}} |f(x)| dx$$

$$\le \|f\|_p (m\{x \in \mathbb{R}^n : |f(x)| > \alpha\})^{\frac{1}{p'}}$$

$$\le \|f\|_p \left(\frac{\|f\|_p}{\alpha}\right)^{\frac{p}{p'}} < \infty,$$

we can know that $f_1 \in L^\infty(\mathbb{R}^n)$ and $f_2 \in L^1(\mathbb{R}^n)$. On the other hand, let

$$k_0 = \|T\|_{w(1,1)}, \quad k_1 = \|T\|_{(\infty,\infty)},$$

the conditions of the theorem give that

$$m(\{x \in \mathbb{R}^n : |T(f_2)(x)| > \alpha\}) \le k_0 \frac{\|f_2\|_1}{\alpha}$$

and

$$\|T(f_1)\|_\infty \le k_1 \|f_1\|_\infty \le k_1 \alpha.$$

Then, we have

$$m(\{x \in \mathbb{R}^n : |T(f)(x)| > 2k_1\alpha\})$$

$$\le m(\{x \in \mathbb{R}^n : |T(f_2)(x)| > k_1\alpha\}) \le \frac{k_0}{k_1} \frac{\|f_2\|_1}{\alpha}.$$

Write $\lambda_g(t)$ as the distribution function of g (see Section 1.3), it holds that

$$\|T(f)\|_p^p = p \int_0^\infty \beta^{p-1} \lambda_{|T(f)|}(\beta) d\beta \leq (2k_1)^p p \int_0^\infty \alpha^{p-1} \lambda_{|T(f)|}(2k_1\alpha) d\alpha,$$

which implies that

$$\|T(f)\|_p^p \leq (2k_1)^p p \frac{k_0}{k_1} \int_0^\infty \alpha^{p-1} \frac{\|f_2\|_1}{\alpha} d\alpha$$

$$= (2k_1)^p p \frac{k_0}{k_1} \int_0^\infty \alpha^{p-2} \left(\int_0^\infty \lambda_{|f_2|}(t) dt \right) d\alpha$$

$$= (2k_1)^p p \frac{k_0}{k_1} \int_0^\infty \alpha^{p-2} \int_0^\infty m(\{x \in \mathbb{R}^n : |f_2(x)| > t\}) dt d\alpha.$$

Noting that $|f_2(x)| > t$ if and only if $|f(x)| > \alpha + t$, we have

$$\|T(f)\|_p^p \leq (2k_1)^p p \frac{k_0}{k_1} \int_0^\infty \alpha^{p-2} \int_\alpha^\infty \lambda_{|f|}(u) du d\alpha$$

$$= (2k_1)^p p \frac{k_0}{k_1} \int_0^\infty \lambda_{|f|}(u) \left(\int_0^u \alpha^{p-2} d\alpha \right) du$$

$$= (2k_1)^p \frac{p}{p-1} \frac{k_0}{k_1} \int_0^\infty u^{p-1} \lambda_{|f|}(u) du$$

$$= 2^p \frac{1}{p-1} k_1^{p-1} k_0 \|f\|_p^p.$$

Therefore,

$$\|T(f)\|_p \leq 2 \left(\frac{1}{p-1} \right)^{\frac{1}{p}} k_0^{\frac{1}{p}} k_1^{1-\frac{1}{p}} \|f\|_p.$$

Note that

$$(p-1)^{p-1} \leq p^p,$$

thus

$$\left(\frac{1}{p-1} \right)^{\frac{1}{p}} \leq \frac{p}{p-1}.$$

This completes the proof of Theorem 4.2.2. □

By inequality

$$\alpha^q(m\{x \in \mathbb{R}^n : |f(x)| > \alpha\}) \leq \int_{\{x \in \mathbb{R}^n : |f(x)| > \alpha\}} |f(x)|^q dx \leq \|f\|_q^q,$$

it is clear that the condition

$$\sup_{\alpha>0} \alpha(m\{x \in \mathbb{R}^n : |f(x)| > \alpha\})^{\frac{1}{q}} < \infty$$

is weaker than $\|f\|_q < \infty$. Therefore, the following definition arises.

Definition 4.2.2. Assume that $1 \leq q \leq \infty$, $X \subset \mathbb{R}^n$. If a measurable function $f(x)$ on X satisfies

$$\sup_{\alpha>0} \alpha(m\{x \in X : |f(x)| > \alpha\})^{\frac{1}{q}} < \infty,$$

then f is said to belong to weak $L^q(X)$, written as $f \in WL^q(X)$ and stipulate that

$$\|f\|_{WL^q(X)} = \sup_{\alpha>0} \alpha(m\{x \in X : |f(x)| > \alpha\})^{\frac{1}{q}},$$

written as $\|f\|_{WL^\infty} = \|f\|_\infty$.

According to this definition, the inequality in Definition 4.0.2 (i) can be abbreviated as

$$\|T(f)\|_{WL^q} \leq C\|f\|_p.$$

Obviously, $L^p(\mathbb{R}^n) \subset WL^p(\mathbb{R}^n)$ and

$$\|f\|_{WL^p} \leq \|f\|_p.$$

Theorem 4.2.3. *Let* $m(X) < \infty$. *If* $1 \leq p_1 < p_2 \leq \infty$, *then* $WL^{p_2}(X) \subset L^{p_1}(X)$.

Proof. Write $\lambda_g(t)$ as the distribution function of g. If $g \in L^p(X)$, $1 \leq p < \infty$, then Corollary 1.3.6 indicates

$$\int_X |g(x)|^p dx = p \int_0^\infty t^{p-1} \lambda_{|g|}(t) dt,$$

where

$$\lambda_{|g|}(t) = m(\{x \in X : |g(x)| > t\}).$$

Let $f \in WL^{p_2}(X)$, that is,

$$\sup_{\alpha > 0} \alpha \{\lambda_{|f|}(\alpha)\}^{\frac{1}{p_2}} = C < \infty,$$

then

$$\int_X |f(x)|^{p_1} dx = p_1 \int_0^\infty t^{p_1-1} \lambda_{|f|}(t) dt$$

$$= p_1 \int_0^1 t^{p_1-1} \lambda_{|f|}(t) dt + p_1 \int_1^\infty t^{p_1-1} \lambda_{|f|}(t) dt.$$

It follows from $\lambda_{|f|}(t) \leq m(X)$ that

$$\int_0^1 t^{p_1-1} \lambda_{|f|}(t) dt < \infty$$

and

$$\int_0^\infty t^{p_1-1} \lambda_{|f|}(t) dt \leq \int_1^\infty t^{p_1-1} \left(\frac{C}{t}\right)^{p_2} dt < \infty.$$

Therefore, $f \in L^{p_1}(X)$, which completes the proof of Theorem 4.2.3. □

Remark. The discussion of the Marcinkiewicz theorem in this section is applicable to quasi-linear operators. The discussion about sublinear operators is just to keep the notation a little simpler.

4.3   Application

Let $K(x, y)$ be a measurable function on $\mathbb{R}^n \times \mathbb{R}^n$. What will be studied is the linear operator T in the form

$$f \mapsto T(f)(x) = \int_{\mathbb{R}^n} K(x, y) f(y) dy,$$

where f is a measurable function with some properties, and the type of the operator T entirely depends on the property of function $K(x, y)$.

Theorem 4.3.1. *Let $1 \le p \le \infty$ if there is a positive number C such that*

$$\int_{\mathbb{R}^n} |K(x, y)| dx \le C, \ a.e. \ y \in \mathbb{R}^n;$$

$$\int_{\mathbb{R}^n} |K(x, y)| dy \le C, \ a.e. \ x \in \mathbb{R}^n,$$

then T is of type (p, p), and

$$\|T\|_{(p,p)} \le C.$$

Proof.   On the one hand, let $f \in L^\infty(\mathbb{R}^n)$, the inequality

$$|T(f)(x)| \le \int_{\mathbb{R}^n} |K(x, y)| dy \|f\|_\infty \le C\|f\|_\infty$$

shows that T is of type (∞, ∞) and $\|T\|_{(\infty,\infty)} \le C$.
On the other hand, let $f \in L(\mathbb{R}^n)$, the inequality

$$\int_{\mathbb{R}^n} |T(f)(x)| dx \le \int_{\mathbb{R}^n} \int_{\mathbb{R}^n} |K(x, y)| dx |f(y)| dy \le C\|f\|_1$$

indicates that T is of type $(1, 1)$ and $\|T\|_{(1,1)} \le C$.
Thus, it follows from the Riesz–Thörin theorem that T is of type (p, p) and $\|T\|_p \le C^{1-t} C^t = C$. The desired conclusion is proved. $\square$

Theorem 4.3.2. *Let $1 < q < \infty$, $\frac{1}{q} + \frac{1}{q'} = 1$ and*

$$\|K(x, \cdot)\|_{WL^q} \le C, \ a.e. \ x \in \mathbb{R}^n,$$

$$\|K(\cdot, y)\|_{WL^q} \le C, \ a.e. \ y \in \mathbb{R}^n.$$

If $1 \le p < q'$, $\frac{1}{p} + \frac{1}{q} = \frac{1}{r} + 1$, then

(1) *T is of weak type $(1, q)$, $p = 1$;*
(2) *T is of type (p, r), $1 < p < q'$.*

Proof. At first, it is worth to point out that if T is of weak type (p, r) for all $1 \le p < q'$, then both (1) and (2) hold. In fact, taking $p = 1$ yields that (1) holds. In order to prove (2), take $\tilde{p}$ such that $p < \tilde{p} < q'$ and denote $\tilde{r} = [\tilde{p}^{-1} - (q')^{-1}]^{-1}$, then $\tilde{p}^{-1} + q^{-1} = \tilde{r}^{-1} + 1$. Therefore, T is of weak type $(\tilde{p}, \tilde{r})$. In the Marcinkiewicz theorem, taking $(p_0, q_0) = (1, q)$, $(p_1, q_1) = (\tilde{p}, \tilde{r})$ implies that T is of type (p_t, q_t), and it is easy to see that there exists a $t \in (0, 1)$ such that $(p_t, q_t) = (p, r)$. Thus, (2) is true.

Next, it will be proved that for every $1 \le p < q'$, T is of weak type (p, r).

Let $A > 0$ (to be determined later). Define

$$E = \{(x, y) : |K(x, y)| > A\},$$

$$K_1(x, y) = \operatorname{sgn}K(x, y)\big[|K(x, y)| - A\big]\chi_E(x, y),$$

$$K_2(x, y) = K(x, y) - K_1(x, y),$$

and

$$T_j(f)(x) = \int_{\mathbb{R}^n} K_j(x, y) f(y) dy, \quad j = 1, 2.$$

Obviously, $T = T_1 + T_2$.

It should be pointed out that K_1 satisfies the conditions of Theorem 4.3.1. In fact, the equality

$$\int_{\mathbb{R}^n} |K_1(x, y)| dx = \int_0^\infty \lambda_{|K_1(\cdot, y)|}(t) dt$$

indicates that for almost every $y \in \mathbb{R}^n$,

$$\int_{\mathbb{R}^n} |K_1(x, y)| dx \le \int_0^\infty \lambda_{|K(\cdot, y)|}(t + A) dt$$

$$\le \|K(\cdot, y)\|_{WL^q}^q \int_A^\infty t^{-q} dt$$

$$\le \frac{C^q A^{1-q}}{q - 1}$$

and

$$\int_{\mathbb{R}^n} |K_1(x,y)|\,dy \le \frac{C^q A^{1-q}}{q-1}, \quad \text{a.e. } x \in \mathbb{R}^n,$$

then it follows from Theorem 4.3.1 that

$$\|T_1(f)\|_p \le \frac{C^q A^{1-q}}{q-1}\|f\|_p.$$

Set $\frac{1}{p} + \frac{1}{p'} = 1$. If $p = 1$, that is, $p' = \infty$, then for any $x \in \mathbb{R}^n$,

$$\|K_2(x,\cdot)\|_{p'} \le A. \tag{4.4}$$

Suppose that $1 < p < q'$, then by the inequality

$$\begin{aligned}
\int_{\mathbb{R}^n} |K_2(x,y)|^{p'}\,dy &= p' \int_0^A t^{p'-1} \lambda_{|K(x,\cdot)|}(t)\,dt \\
&\le \|K(x,\cdot)\|_{WL^q}^q\, p' \int_0^A t^{p'-1-q}\,dt \\
&\le \frac{p' C^q}{p'-q} A^{p'-q}, \quad \text{a.e. } x \in \mathbb{R}^n,
\end{aligned}$$

and the fact $\frac{1}{p} + \frac{1}{q} = \frac{1}{r} + 1$, one may get

$$\|K_2(x,\cdot)\|_{p'} \le \left(\frac{r}{q}C^q\right)^{\frac{1}{p'}} A^{\frac{q}{r}}, \quad \text{a.e. } x \in \mathbb{R}^n. \tag{4.5}$$

Obviously, (4.4) can be merged into (4.5), then by Hölder's inequality,

$$\|T_2(f)\|_\infty \le \left(\frac{r}{q}C^q\right)^{\frac{1}{p'}} A^{\frac{q}{r}}\|f\|_p.$$

Thus, taking A satisfying

$$\left(C^q \frac{r}{q}\right)^{\frac{1}{p'}} A^{\frac{q}{r}}\|f\|_p = \frac{t}{2},$$

we can obtain $\lambda_{|T_2(f)|}\left(\frac{t}{2}\right) = 0$, therefore

$$\lambda_{|T(f)|}(t) \le \lambda_{|T_1(f)|}\left(\frac{t}{2}\right) \le (2\|T_1(f)\|_p t^{-1})^p$$

$$\le \left(\frac{2C^q}{q-1}A^{1-q}\|f\|_p t^{-1}\right)^p$$

$$\le \left[\frac{2C^q}{q-1}(\frac{r}{q})^{\frac{r}{p'q'}}\left(\frac{1}{2C^{\frac{q}{p'}}}\frac{t}{\|f\|_p}\right)^{\frac{r(1-q)}{q}}\frac{\|f\|_p}{t}\right]^p$$

$$= \left(\frac{1}{q-1}\right)^p\left(\frac{r}{q}\right)^{\frac{rp}{p'q'}}(2C\|f\|_p t^{-1})^r,$$

which indicates that T is of weak type (p,r), $1 \le p < q'$. $\qquad\square$

Exercise 4

1. Let $1 \le r, p, q \le \infty$, $a(x) \in L^r(\mathbb{R}^n)$. Define the operator
$$T: f \mapsto Tf = af.$$
Prove that T is of type (p, q) when
$$\frac{1}{q} = \frac{1}{r} + \frac{1}{p}.$$

2. Suppose that $0 < \alpha < n$. Define the operator T_α:
$$T_\alpha(f)(x) = \int_{\mathbb{R}^n} |x - y|^{-\alpha} f(y) dy.$$
Prove that T_α is of weak type $(1, \frac{n}{\alpha})$.

3. Let the operator T be of weak type (p, q) and $1 \le p, q < \infty$. Prove that if $m(X) < \infty$, $0 < r < q$, then
$$\left(\int_X |T(f)(x)|^r dx\right)^{\frac{1}{r}} \le C\left(\frac{q}{q-r}\right)^{\frac{1}{r}} m(X)^{\frac{1}{r}-\frac{1}{q}}\|f\|_p,$$
where $C = \|T\|_{w(p,q)}$.
   (*Hint*: Take $\lambda = C\frac{\|f\|_p}{m(X)^{\frac{1}{q}}}$ and divide f into two parts according to the level λ.)

4. Use Hölder's inequality to prove Theorem 4.3.1.

Chapter 5

Hardy–Littlewood Maximal Function

5.1 The Lebesgue Differentiation Theorem

For a one-variable integrable function f, the function F of the form $F(x) = \int_a^x f(t)dt$, where a is an arbitrary constant, is called the original function of f or antiderivative. We try to extend this concept to the multivariate case in a certain sense.

Let f be a measurable function on $\mathbb{R}^n$, and the integral value $\int_{\mathbb{R}^n} f(x)dx$ exists. For any Lebesgue measurable set $E \subset \mathbb{R}^n$, define $F(E) = \int_E f(x)dx$ and the set function F is called the indefinite integral of f. Unlike the one-variable case, the indefinite integral (when $n > 1$) cannot be defined as a point function, but as a set function, or rather, it is a signed measure, which is absolutely continuous with respect to the Lebesgue measure, and the Radon–Nikodym derivative is the original function f.

Let $x = (x_1, \ldots, x_n) \in \mathbb{R}^n$, $r > 0$, and write

$$Q(x,r) = \left\{ y \in \mathbb{R}^n : x_j - \frac{r}{2} < y_j < x_j + \frac{r}{2}, j = 1, \ldots, n \right\}.$$

That is, $Q(x,r)$ denotes an open cube with the center x, the side length r and each side is parallel to each coordinate axis. The side length r will be omitted and $Q(x,r)$ will be written as $Q(x)$ when the side length does not need to be noticed; we omit the center x and simply write $Q(x,r)$ as $Q(r)$ when the center does not need to be noticed. Of course, these abbreviations are in no way confusing as long as they are contextualized.

Definition 5.1.1. Given $x \in \mathbb{R}^n$, the indefinite integral of f is said to be derivable at x and its derivative is $f(x)$ if

$$\lim_{r \to 0} r^{-n} F(Q(x, r)) = f(x).$$

Theorem 5.1.1. *Let $f \in C_c(\mathbb{R}^n)$, then the indefinite integral of f is derivable everywhere on $\mathbb{R}^n$.*

Proof. Take $x \in \mathbb{R}^n$, $r > 0$. The inequality

$$\left| r^{-n} F(Q(x, r)) - f(x) \right| \leq r^{-n} \int_{Q(x,r)} |f(t) - f(x)| dt$$

$$\leq \sup_{y \in Q(x,r)} |f(y) - f(x)|$$

and the uniform continuity of f give

$$\lim_{r \to 0} r^{-n} F(Q(x, r)) = f(x).$$

$\square$

The following theorem is called the Lebesgue differentiation theorem.

Theorem 5.1.2. *If $f \in L(\mathbb{R}^n)$, then the indefinite integral of f is derivable almost everywhere on $\mathbb{R}^n$, and*

$$\lim_{r \to 0} r^{-n} F(Q(x, r)) = f(x), \quad \text{a.e.}$$

In order to prove this theorem, we quote Theorem 2.4.2 to obtain the following lemma.

Lemma 5.1.3. *Let $f \in L^p(\mathbb{R}^n)$, $1 \leq p < \infty$, then for any $\varepsilon > 0$, there exists $C \in C_c(\mathbb{R}^n)$ such that*

$$\int_{\mathbb{R}^n} |f(x) - C(x)|^p dx < \varepsilon.$$

Obviously, it suffices to show that

$$\limsup_{r \to 0} \left| r^{-n} F(Q(x, r)) - f(x) \right| = 0, \quad \text{a.e.}$$

It follows from Lemma 5.1.3 that there exists $C_k \in C_C(\mathbb{R}^n)$ satisfying

$$\int_{\mathbb{R}^n} |f(t) - C_k(t)| dt \to 0, \quad k \to \infty,$$

then by Theorem 5.1.1, for any fixed k,

$$\limsup_{r \to 0} |r^{-n} F(Q(x,r)) - f(x)|$$

$$\leq \limsup_{r \to 0} \left| r^{-n} F(Q(x,r)) - r^{-n} \int_{Q(x,r)} C_k(t) dt \right|$$

$$+ \limsup_{r \to 0} \left| r^{-n} \int_{Q(x,r)} C_k(y) dy - C_k(x) \right| + |C_k(x) - f(x)|$$

$$\leq \sup_{r > 0} r^{-n} \int_{Q(x,r)} |f(y) - C_k(y)| dy + |C_k(x) - f(x)|.$$

From the right-hand side of the above inequality, we need to deal with a class of functions of the form

$$x \mapsto \sup_{r > 0} r^{-n} \int_{Q(x,r)} |g(y)| dy,$$

called Hardy–Littlewood (*HL*) maximal function of g or abbreviated as the *HL* maximal function. The properties of the *HL* maximal function will be established in Section 5.3, and Theorem 5.1.2 will be proved in Section 5.3.

5.2 The Covering Lemma

Lemma 5.2.1 (Vitali's covering lemma). *Let $m(E) < \infty$, then a finite number of disjoint cubes $\{Q_j\}_{j=1}^N$ can be selected from the cube covering $\mathcal{K}$ of E such that*

$$\sum_{j=1}^N |Q_j| \geq 5^{-n-1} m(E).$$

Proof. Let $Q = Q(t)$ denote that the side length of the cube Q is t and $\mathcal{K}_1 = \mathcal{K}$. Define

$$t_1^* = \sup \{t : Q = Q(t) \in \mathcal{K}_1\}.$$

Assume that $t_1^* < \infty$, or there is nothing to prove. Then, take $Q_1 = Q_1(t_1) \in \mathcal{K}_1$ such that $t_1 > \frac{t_1^*}{2}$. Set $\mathcal{K}_1 = \mathcal{K}_2 \cup \mathcal{K}_2'$, where

$$\mathcal{K}_2 = \{Q \in \mathcal{K}_1 : Q \cap Q_1 = \emptyset\},$$
$$\mathcal{K}_2' = \{Q \in \mathcal{K}_1 : Q \cap Q_1 \neq \emptyset\}.$$

Let Q_1^5 denote the cube with the same center as Q whose side length is $5Q_1$. The fact $t_1^* < 2t_1$ shows that for any $Q \in \mathcal{K}_2'$, $Q \subset Q_1^5$. Also, define

$$t_2^* = \sup\{t : Q = Q(t) \in \mathcal{K}_2\}.$$

Similarly, one can take $Q_2 = Q_2(t_2) \in \mathcal{K}_2$ such that $t_2 > \frac{t_2^*}{2}$. Clearly, $Q_1 \cap Q_2 = \emptyset$.

Set $\mathcal{K}_2 = \mathcal{K}_3 \cup \mathcal{K}_3'$, where

$$\mathcal{K}_3 = \{Q \in \mathcal{K}_2 : Q \cap Q_2 = \emptyset\},$$
$$\mathcal{K}_3' = \{Q \in \mathcal{K}_2 : Q \cap Q_2 \neq \emptyset\}.$$

Likewise, for any $Q \in \mathcal{K}_3'$, $Q \subset Q_2^5$ holds. Repeating the above process can obtain a sequence of cubes $\{Q_i\}$. Obviously, all Q_i are disjoint, and each $Q \in \mathcal{K}_{i+1}'$ satisfies $Q \subset Q_i^5$. In the following, two cases are possible:

(i) At a certain step, $\mathcal{K}_{N+1} = \emptyset$. In this case, the resulting sequence of cubes is finite, and $\mathcal{K} = \mathcal{K}_2' \cup \mathcal{K}_3' \cup \cdots \cup \mathcal{K}_{N+1}'$, then

$$m(E) \leq m\left(\bigcup_{i=1}^N \bigcup_{Q \in \mathcal{K}_{i+1}'} Q\right) \leq m\left(\bigcup_{i=1}^N Q_i^5\right) = 5^n \sum_{i=1}^N m(Q_i).$$

(ii) Otherwise, the resulting sequence of cubes $\{Q_i\}$ is infinite. Let $\lim_{j\to\infty} t_j^* = 0$ (or else, by $t_j^* \searrow \delta$, $t_j \geq \frac{1}{2}t_j^* \geq \frac{1}{2}\delta > 0$ and $\sum_{j=1}^\infty m(Q_j(t_j)) = \infty$, the conclusion of the lemma holds). In this case, for any $Q = Q(t) \in \mathcal{K} = \mathcal{K}_1$, we must have $Q \subset \bigcup_j Q_j^5$. Indeed, if $Q \not\subset \bigcup_j Q_j^5$, then $Q \notin \mathcal{K}_{j+1}'$ for every $j \in \mathbb{N}$. Hence, $Q \in \mathcal{K}_{j+1}$ for every $j \in \mathbb{N}$. The fact that $t \leq t_{j+1}^*$ and $\lim_{j\to\infty} t_j^* = 0$ leads to the false conclusion $t = 0$. Therefore,

$$E \subset \bigcup_{Q \in \mathcal{K}} Q \subset \bigcup_j Q_j^5,$$

which implies that

$$m(E) \le m\left(\bigcup_j Q_j^5\right) = 5^n \sum_{j=1}^{\infty} m(Q_j).$$

$\square$

Remark. The conclusion still holds if the cubes in the lemma are replaced by balls.

5.3 Maximal Function *HL*

Definition 5.3.1. Let $f \in L_{\text{loc}}(\mathbb{R}^n)$ (this notation denotes the set of locally integrable functions which are integrable on every compact subset of $\mathbb{R}^n$). The function

$$HL(f)(x) = \sup_{r>0} r^{-n} \int_{Q(x,r)} |f(y)| dy, \quad x \in \mathbb{R}^n,$$

is called the Hardy–Littlewood maximal function of f and HL is the Hardy–Littlewood maximal operator.

Theorem 5.3.1. *The operator* $HL : f \mapsto HL(f)$ *is of weak type* $(1, 1)$, *and for any* $\alpha > 0$,

$$m(\{x \in \mathbb{R}^n : HL(f)(x) > \alpha\}) \le 5^{n+1}\alpha^{-1}\|f\|_1.$$

Proof. Let $\alpha > 0$, $k \in \mathbb{N}$. Define

$$E = \{x \in \mathbb{R}^n : HL(f)(x) > \alpha\}, \quad E_k = \{x \in E : |x| < k\},$$

where $|x| = (x_1^2 + \cdots + x_n^2)^{\frac{1}{2}}$ is the Euclid norm of x. Clearly,

$$mE_k < \infty, \quad \lim_{k\to\infty} mE_k = mE.$$

For each $x \in E_k$, there exists a cube $Q(x)$ such that

$$\frac{1}{mQ(x)} \int_{Q(x)} |f(y)| dy > \alpha.$$

Obviously, $\{Q(x) : x \in E_k\}$ is a cover of E_k. It follows from Lemma 5.2.1 that a finite number of $Q_j : j = 1, \ldots, N$, which are pairwise disjoint and satisfy

$$mE_k \leq 5^{n+1} \sum_{j=1}^{N} |Q_j|,$$

can be extracted from this family of cubes. Thus,

$$mE_k \leq 5^{n+1} \sum_{j=1}^{N} \frac{1}{\alpha} \int_{Q_j} |f(y)| dy \leq 5^{n+1} \alpha^{-1} \|f\|_1,$$

the proof of the theorem follows by letting $k \to \infty$. $\square$

Apparently, the sub-linear operator $: f \mapsto HL(f)$ is of type (∞, ∞). Combining with Theorems 4.2.2 and 5.3.1 yields the following.

Corollary 5.3.2. *Let $1 < p < \infty$, then the operator HL is of type (p, p) and*

$$\|HL(f)\|_p \leq C \frac{p}{p-1} \|f\|_p,$$

where C is independent of p and f.

Let us go back to the proof of the Lebesgue differentiation theorem.

Proof.   It was pointed out at the end of Section 5.1 that

$$\limsup_{r \to 0} |r^{-n} F(Q(x,r)) - f(x)| \leq HL(f - C_k)(x) + |C_k(x) - f(x)|,$$

where $C_k \in C_c(\mathbb{R}^n)$ and satisfies

$$\int_{\mathbb{R}^n} |f(t) - C_k(t)| dt \to 0, \quad k \to \infty.$$

For any $\varepsilon > 0$, denote

$$E_\varepsilon = \left\{ x \in \mathbb{R}^n : \limsup_{r \to 0} |r^{-n} F(Q(x,r)) - f(x)| > \varepsilon \right\}.$$

Note that

$$E_\varepsilon \subset \left\{ x \in \mathbb{R}^n : HL(f - C_k)(x) > \frac{\varepsilon}{2} \right\}$$

$$\bigcup \left\{ x \in \mathbb{R}^n : |f(x) - C_k(x)| > \frac{\varepsilon}{2} \right\},$$

then by Theorem 5.3.1, we derive

$$mE_\varepsilon \le 5^{n+1}\frac{2}{\varepsilon}\int_{\mathbb{R}^n}|f(y)-C_k(y)|dy + \frac{2}{\varepsilon}\int_{\mathbb{R}^n}|f(y)-C_k(y)|dy,$$

which implies that $mE_\varepsilon = 0$. $\qquad\square$

Remark. It is clear that the condition $f \in L(\mathbb{R}^n)$ in the Lebesgue differentiation theorem can be relaxed to $f \in L_{\text{loc}}(\mathbb{R}^n)$.

As a simple application of the maximal function, a stronger conclusion than the Lebesgue differentiation theorem will be presented in the following.

Definition 5.3.2. Let $1 \le p < \infty$, $f \in L^p(\mathbb{R}^n)$, the point x is called the Lebesgue point of f with power p, abbreviated as L^p point if

$$\lim_{r\to 0} r^{-n}\int_{Q(x,r)}|f(y)-f(x)|^p dy = 0$$

holds at x, and the Lebesgue point with power 1 is abbreviated as Lebesgue point or L point.

Obviously, the concept of Lebesgue point is to characterize local properties, which can be defined only for locally integrable (or locally p-power integrable) functions. Moreover, L^p points, $1 < p < \infty$, must be L points.

Theorem 5.3.3. *Let $1 \le p < \infty$, $f \in L^p(\mathbb{R}^n)$, then*

$$\lim_{r\to 0} r^{-n}\int_{Q(x,r)}|f(y)-f(x)|^p dy = 0, \quad a.e. \ x \in \mathbb{R}^n,$$

that is, almost all points on $\mathbb{R}^n$ are the Lebesgue points of f with power p.

Proof. Let $\{r_k : k \in \mathbb{N}\}$ be the set of rational numbers in $\mathbb{R}$ and write

$$Z_k = \left\{ x \in \mathbb{R}^n : \lim_{r\to 0} r^{-n}\int_{Q(x,r)}|f(y)-r_k|^p dy \ne |f(x)-r_k|^p \right\}.$$

By $|f(y) - r_k|^p \in L_{\text{loc}}(\mathbb{R}^n)$ and Remark above, we have for $k \in \mathbb{N}$, $m(Z_k) = 0$.

Denote

$$Z_0 = \{x \in \mathbb{R}^n : |f(x)| = \infty\}, \quad Z = \bigcup_{k=0}^{\infty} Z_k.$$

Note that $m(Z) = 0$, then it suffices to prove that when $x \notin Z$,

$$\limsup_{r \to 0} r^{-n} \int_{Q(x,r)} |f(y) - f(x)|^p dy = 0.$$

In fact, for any rational number r_k and positive number r,

$$\left(r^{-n} \int_{Q(x,r)} |f(y) - f(x)|^p dy \right)^{\frac{1}{p}}$$

$$\leq \left(r^{-n} \int_{Q(x,r)} |f(y) - r_k|^p dy \right)^{\frac{1}{p}} + |f(x) - r_k|,$$

which yields that when $x \notin Z$,

$$\limsup_{r \to 0} \left(r^{-n} \int_{Q(x,r)} |f(y) - f(x)|^p dy \right)^{\frac{1}{p}} \leq 2|f(x) - r_k|.$$

The right-hand side of the inequality above is arbitrarily small, which is a confirmation of the previous statement. □

By the way, the operator HL is not of type $(1, 1)$. The following example shows that $f \in L(\mathbb{R}^n)$ does not guarantee that $HL(f) \in L(\mathbb{R}^n)$.

Example 5.3.1. Let $Q = Q(O, 1)$, where O is the origin, then $HL(\chi_Q) \notin L(\mathbb{R}^n)$.

Proof. In fact, when $|x| > 2$,

$$HL(\chi_Q)(x) = \sup_{r>0} r^{-n} \int_{Q(x,r)} \chi_Q(y) dy$$

$$= \sup_{r>0} r^{-n} m(Q \cap Q(x,r))$$

$$\geq (4|x|)^{-n} m\big(Q \cap Q(x, 4|x|)\big)$$
$$\geq (4|x|)^{-n},$$

which indicates that $HL(\chi_Q) \notin L(\mathbb{R}^n)$. $\qquad\square$

As an application of Corollary 5.3.2, $f \in L\log^+ L(X)$ can guarantee $HL(f) \in L(X)$ if X is a finite measure subset of $\mathbb{R}^n$. Precisely, the following theorem holds.

Theorem 5.3.4. *Let $1 < m(X) < \infty$. If*

$$\int_X |f(x)|(1 + \log^+ |f(x)|)dx < \infty,$$

then $HL(f) \in L(X)$ and

$$\int_X HL(f)(x)dx \leq Cm(X)\left\{1 + \int_X |f(x)|(1 + \log^+ |f(x)|)dx\right\},$$

where $\log^+ |u| = \max\{0, \log|u|\}$ and C is a constant depending on n.

Proof. Write

$$E_0 = \{x \in X : 0 \leq |f(x)| < 1\},$$
$$E_k = \{x \in X : 2^{k-1} \leq |f(x)| < 2^k\}, \quad k \in \mathbb{N}.$$

Denote $f_k = |f|\chi_{E_k}$. It is clear that $|f| = \sum_{k=0}^{\infty} f_k$, a.e. and

$$HL(f)(x) \leq \sum_{k=0}^{\infty} HL(f_k)(x) \leq 1 + \sum_{k=1}^{\infty} HL(f_k)(x).$$

Take $p_k = 1 + \frac{1}{k+1}$, $p_k' = k+2$. Hölder's inequality for the exponential pair (p_k, p_k') yields

$$\int_X HL(f_k)(x)dx \leq \|HL(f_k)\|_{p_k}(m(X))^{\frac{1}{p_k'}}, \quad k \in \mathbb{N},$$

then

$$\int_X HL(f)(x)dx \leq m(X) + \sum_{k=1}^{\infty} \|HL(f_k)\|_{p_k} (m(X))^{\frac{1}{p_k'}}$$

$$\leq m(X) + C\sum_{k=1}^{\infty} \frac{p_k}{p_k - 1} \|f_k\|_{p_k} (m(X))^{\frac{1}{p_k'}}$$

$$\leq m(X) + 2C\sum_{k=1}^{\infty} (k+2)2^k (m(E_k))^{\frac{1}{p_k}} (m(X))^{\frac{1}{p_k'}}$$

$$\leq Cm(X)\left\{1 + 2C\left(\sum_{m(E_k)\leq 3^{-k}} + \sum_{m(E_k)>3^{-k}}\right)\right.$$

$$\left.(k+2)2^k (m(E_k))^{\frac{1}{p_k}} (m(X))^{\frac{1}{p_k'}}\right\}$$

$$\leq Cm(X)\left\{1 + \sum_{k=1}^{\infty} k2^k m(E_k)\right\}.$$

On the other hand, it is clear that

$$\int_X |f(x)|(1 + \log^+ |f(x)|)dx \leq \sum_{k=1}^{\infty} \int_{E_k} |f(x)|(1 + \log^+ |f(x)|)dx.$$

Note that

$$\sum_{k=1}^{\infty} \int_{E_k} |f(x)|(1 + \log^+ |f(x)|)dx \geq \sum_{k=1}^{\infty} 2^{k-1}[1 + (k-1)\log 2]m(E_k).$$

Combining the above equations, the conclusion of the theorem is obtained. □

Exercise 5

1. For $f \in L_{\text{loc}}(\mathbb{R}^n)$ and a cube Q of $\mathbb{R}^n$, define

$$f^*(x) = \sup_{Q \ni x} \frac{1}{m(Q)} \int_Q |f(y)|dy.$$

Prove that there exists a positive constant C (only depending on n) such that

$$Cf^*(x) \leq HL(f)(x) \leq f^*(x).$$

2. Prove that for $f \in L(\mathbb{R}^n)$, $HL(f) \notin L(\mathbb{R}^n)$ if $\|f\|_1 > 0$.
3. Let f be a function on $\mathbb{R}^n$ with period 1 for each variable and

$$Q = Q(O, 1) = \left\{ x = (x_1, \ldots, x_n) : |x_i| < \frac{1}{2},\ 1 \leq i \leq n \right\}.$$

Prove that if $f \in L(Q)$ and $f_Q = f\chi_Q$, then when $2x \in Q$,

$$HL(f)(x) \leq CHL(f_Q)(x),$$

where C is a constant independent of f and x.
4. Prove that for $f \in L_{\mathrm{loc}}(\mathbb{R}^n)$,

$$HL(f)(x) \leq C|x|^{-n} \log |x|$$

if $|f(x)| \leq C|x|^{-n}$ when $|x| > 2$.
5. Let Q be an open cube. Prove that $HL(\chi_Q)(x) = 1$ for any $x \in Q$.
6. Assume that x_0 is the center of a cube Q. For $x \in Q$, prove that there exists a constant C (independent of f, x and Q) such that

$$\left| \int_{\mathbb{R}^n \setminus Q} \frac{|x_0 - x|}{|x_0 - y|^{n+1}} f(y) dy \right| \leq CHL(f)(x).$$

Chapter 6

Convolution

6.1 Convolution

The notion of convolution plays an important role in both classical analysis and modern analysis, see [4].

Definition 6.1.1. Let f and g be measurable functions, the convolution of f and g is defined by

$$(f * g)(x) = \int_{\mathbb{R}^n} f(x - t)g(t)dt.$$

Convolution, regarded as an algebraic operation, satisfies the commutative and associative laws:

$$(f * g)(x) = (g * f)(x),$$

$$(f * g) * h(x) = f * (g * h)(x).$$

In the following, we discuss the three properties of convolution, namely, integrability, continuity and smoothness.

Regarding the integrability of convolution, we have the following theorem.

Theorem 6.1.1. *Let*

$$1 \le p, q, r \le \infty, \quad \frac{1}{p} + \frac{1}{q} = \frac{1}{r} + 1,$$

then for all $f \in L^p(\mathbb{R}^n)$ and $g \in L^q(\mathbb{R}^n)$, we have $f * g \in L^r(\mathbb{R}^n)$ and

$$\|f * g\|_r \le \|f\|_p \|g\|_q.$$

Proof. We start with the case $q = 1$, then $p = r$,

$$f * g(x) = \int_{\mathbb{R}^n} f(x - t)g(t)dt.$$

By Minkowski's inequality (see Theorem 3.4.3), then

$$\|f * g\|_p \leq \int_{\mathbb{R}^n} \|f\|_p \, |g(t)|dt = \|f\|_p \|g\|_1.$$

Fix $f \in L^p(\mathbb{R}^n)$, consider the linear operator $T : g \mapsto f * g$, then the above results indicate that T is of type $(1, p)$, and

$$\|Tg\|_p \leq \|f\|_p \|g\|_1.$$

By Hölder's inequality, it's easy to show that T is of type (p', ∞), and

$$\|Tg\|_\infty \leq \|f\|_p \|g\|_{p'}.$$

Setting $(p_0, q_0) = (1, p)$, $(p_1, q_1) = (p', \infty)$ and interpolating at $t = \frac{p}{q'}$ in the Riesz–Thörin theorem, we can get $p_t = q$, $q_t = r$, and

$$\|f * g\|_r = \|Tg\|_{q_t} \leq (\|f\|_p^{1-t} \|f\|_p^{t}) \|g\|_{p_t} = \|f\|_p \|g\|_q. \qquad \square$$

Before we discuss the continuity of convolution, we introduce the concept of continuous mode. Let f be a uniformly continuous function on $\mathbb{R}^n$. The continuous mode of f is defined as follows:

$$\omega(f; \delta)_c = \sup\{|f(x) - f(y)| : x, y \in \mathbb{R}^n, |x - y| \leq \delta\}, \quad \delta > 0,$$

then clearly

$$\lim_{\delta \to 0} \omega(f; \delta)_c = 0,$$

which is the characterization of uniform continuity for f. For a continuous function f, write $\|f\|_c = \|f\|_\infty$.

For the continuity of convolution, we have the following theorem.

Theorem 6.1.2. *Let $f \in L(\mathbb{R}^n)$, and K be a bounded and uniformly continuous function in $\mathbb{R}^n$, then $f * K$ is bounded and uniformly continuous on $\mathbb{R}^n$.*

Proof. The boundedness of $f * K$ is obvious and the uniform continuity is given by

$$|(f * K)(x + h) - (f * K)(x)| = \left| \int_{\mathbb{R}^n} f(x - t)[K(t + h) - K(t)]dt \right|$$

$$\leq \omega(K; |h|)_c \|f\|_1. \qquad \square$$

A meaningful property of convolution transformation is that the convolution of an integrable function with a smooth function is a smooth function. This fact would make convolution one of the important roles in analysis. Let $m \in \mathbb{Z}_+$, $\alpha = (\alpha_1, \ldots, \alpha_n) \in \mathbb{Z}_+^n$, write

$$C^m(\mathbb{R}^n) = \{f : D^\alpha f \in C(\mathbb{R}^n), |\alpha| \leq m\},$$

$$C_c^m(\mathbb{R}^n) = \{f \in C^m(\mathbb{R}^n) : f \text{ is supported in a compact set}\},$$

where

$$D^\alpha = \frac{\partial^{\alpha_1 + \cdots + \alpha_n}}{\partial x_1^{\alpha_1} \cdots \partial x_n^{\alpha_n}}, \quad |\alpha| = \alpha_1 + \cdots + \alpha_n.$$

Theorem 6.1.3. *Let $K \in C_c^m(\mathbb{R}^n)$, $f \in L_{\mathrm{loc}}(\mathbb{R}^n)$, then $f * K \in C^m(\mathbb{R}^n)$ and*

$$D^\alpha(f * K)(x) = (f * D^\alpha K)(x), \quad |\alpha| \leq m.$$

Proof. Let

$$r = \sup\{|y| : y \in \operatorname{supp} K\}.$$

For any $x \in \mathbb{R}^n$, define

$$B(x) = \{y \in \mathbb{R}^n : |y| \leq r + |x| + 1\},$$

it's obvious that when $t \in \complement B(x)$, $h \in \mathbb{R}^n$ and $|h| < 1$,

$$K(x + h - t) = K(x - t) = 0.$$

Consider the case $m = 0$. Then $f * K \in C(\mathbb{R}^n)$ since

$$|(f * K)(x + h) - (f * K)(x)|$$

$$= \left| \int_{\mathbb{R}^n} f(t)[K(x + h - t) - K(x - t)]dt \right|$$

$$\leq \|f\|_{L(B(x))} \omega(K; |h|)_c, \quad h \in \mathbb{R}^n.$$

Consider the case $m = 1$. Write $h = (0, \ldots, h_i, 0, \ldots, 0)$ (where h_i is ith element). Obviously, when $|h| < 1$,

$$\frac{(f * K)(x + h) - (f * K)(x)}{h_i}$$

$$= \int_{B(x)} f(t) \left[\frac{K(x - t + h) - K(x - t)}{h_i} \right] dt$$

$$= \int_{B(x)} f(t) \frac{\partial K}{\partial x_i}(x - t + h') dt,$$

where $h' = (0, \ldots, h'_i, 0, \ldots, 0)$ (h'_i is ith element and $0 < |h'_i| < |h_i|$). The same as the case $m = 0$,

$$\left| \frac{(f * K)(x + h) - (f * K)(x)}{h_i} - f * \frac{\partial K}{\partial x_i} \right| \leq \|f\|_{L(B(x))} \omega \left(\frac{\partial K}{\partial x_i}; |h| \right)_c.$$

Then let $h_i \to 0$, we have

$$\frac{\partial}{\partial x_i}(f * K)(x) = \left(f * \frac{\partial K}{\partial x_i} \right)(x).$$

For the cases $m = 2, 3, \ldots$, we can prove them in the same way. $\square$

The following corollary can be deduced by Theorem 6.1.3 directly.

Corollary 6.1.4.

(i) *Let $f \in L_{\mathrm{loc}}(\mathbb{R}^n)$ and $K \in C_c^\infty(\mathbb{R}^n)$, then*

$$f * K \in C^\infty(\mathbb{R}^n).$$

(ii) *Let f be a function in $L(\mathbb{R}^n)$ that is supported in a compact set, and $K \in C_c^\infty(\mathbb{R}^n)$, then*

$$f * K \in C_c^\infty(\mathbb{R}^n).$$

Now, we use Corollary 6.1.4(ii) to give a specific method of constructing the function f in the Uryson lemma. This, of course, is based on the Euclidean topology of $\mathbb{R}^n$.

Theorem 6.1.5 (Uryson's lemma on $\mathbb{R}^n$). *Let K be a compact subset of $\mathbb{R}^n$, U be an open subset and $K \subset U$, then there exists a* *function $f \in C_c^\infty(\mathbb{R}^n) : \mathbb{R}^n \to [0,1]$, such that*

$$f(x) = \begin{cases} 1, & x \in K, \\ 0, & x \in \complement U. \end{cases}$$

Proof. Write $\delta = dist(K, \complement U)$. Since K is a compact subset of $\mathbb{R}^n$, it follows that $\delta > 0$. Define

$$V = \left\{ x : dist(x, K) < \frac{\delta}{3} \right\}.$$

Now, fix a nonnegative function $\psi \in C_c^\infty(\mathbb{R}_+)$ defined on $\mathbb{R}_+ = [0, \infty)$ satisfying

$$\text{supp } \psi \subset \left[0, \frac{\delta}{3}\right), \quad \left(\int_0^\infty t^{n-1}\psi(t)dt\right) \sigma(\Sigma_{n-1}) = 1.$$

Denote

$$\varphi(x) = \psi(|x|), \quad x \in \mathbb{R}^n,$$

then we can deduce that $\varphi \in C_c^\infty(\mathbb{R}^n)$ and

$$\text{supp } \varphi \subset \left\{ x : |x| < \frac{\delta}{3} \right\}, \quad \int_{\mathbb{R}^n} \varphi(x)dx = 1.$$

Let $f = \chi_V * \varphi$, Corollary 6.1.4(ii) implies that $f \in C_c^\infty(\mathbb{R}^n)$. Obviously, $0 \le f \le 1$. When $x \in K$, and $t \in \text{supp } \varphi$, we have $x - t \in V$. Then

$$f(x) = \int_{\mathbb{R}^n} \chi_V(x-t)\varphi(t)dt = \int_{\mathbb{R}^n} \varphi(t)dt = 1, \quad x \in K.$$

It's easy to get supp $f \subset U$ from supp $\chi_V * \varphi \subset$ supp $\chi_V +$ supp φ. $\qquad \square$

6.2 Approximate Identity

We still use $\mathscr{F}$ to represent the set of measurable functions on $\mathbb{R}^n$.

Definition 6.2.1. Let K be a measurable function on $\mathbb{R}^n$, $K_\varepsilon(x) = \varepsilon^{-n}K(\frac{x}{\varepsilon})$ with $\varepsilon > 0$ and X be a subset of $\mathscr{F}$. Then K is said to be a kernel of approximate identity on X if for any f on X (in a certain sense of convergence), one can get

$$f * K_\varepsilon \to f, \quad \text{when } \varepsilon \to 0,$$

the operator $K_\varepsilon : f \mapsto f * K_\varepsilon$ is an approximate identity operator on X (when $\varepsilon \to 0$).

We first study the conditions under which K_ε becomes an approximate identity operator in the sense of norm convergence. The following two theorems show that, in principle, as long as K is integrable, it's an approximate identity kernel on the scale $L^p(1 \le p < \infty)$ and on the uniform scale.

Theorem 6.2.1. *Let $K \in L(\mathbb{R}^n)$ and $\int_{\mathbb{R}^n} K(x)dx = 1$. If $f \in L^p(\mathbb{R}^n)$, $1 \le p < \infty$, then*

$$\lim_{\varepsilon \to 0} \|f * K_\varepsilon - f\|_p = 0.$$

Proof. Minkowski's inequality (see Theorem 3.4.3) yields that

$$\|f * K_\varepsilon - f\|_p \le \left| \int_{\mathbb{R}^n} \|f(\cdot - \varepsilon t) - f(\cdot)\|_p \, |K(t)|dt \right|.$$

Combining

$$\|f(\cdot - \varepsilon t) - f(\cdot)\|_p \le 2\|f\|_p$$

and

$$\lim_{\varepsilon \to 0} \|f(\cdot - \varepsilon t) - f(\cdot)\|_p = 0,$$

which can be deduced by Lemma 5.1.3, with the Lebesgue dominated convergence theorem, we finish our proof. $\qquad\square$

Theorem 6.2.2. *Let $K \in L(\mathbb{R}^n)$ and $\int_{\mathbb{R}^n} K(x)dx = 1$. If f is bounded and uniformly continuous, then*

$$\lim_{\varepsilon \to 0} \|f * K_\varepsilon - f\|_c = 0.$$

Proof. We have

$$|(f * K_\varepsilon)(x) - f(x)| = \left| \int_{\mathbb{R}^n} [f(x - \varepsilon t) - f(x)]K(t)dt \right|$$

$$\le \int_{\mathbb{R}^n} \omega(f; \varepsilon|t|)_c \, |K(t)|dt.$$

By

$$\omega(f; \varepsilon |t|)_c \le 2\|f\|_c,$$

and the Lebesgue dominated convergence theorem, the conclusion can be proved. $\qquad\square$

As a corollary of Theorem 6.2.1, we get Theorem 2.4.2 again and reinforce the conclusion of Lemma 5.1.3 as follows.

Corollary 6.2.3. *Let* $1 \le p < \infty$, *then* $C_c^\infty(\mathbb{R}^n)$ *is dense in* $L^p(\mathbb{R}^n)$.

Proof. Assume $f \in L^p(\mathbb{R}^n)$. Let $f = g + h$, where g has a compact supported set and $\|h\|_p < \eta$. Take $K \in C_c^\infty(\mathbb{R}^n)$ and $\int_{\mathbb{R}^n} K(x) = 1$. We can get $g * K_\varepsilon \in C_c^\infty(\mathbb{R}^n)$ by Corollary 6.1.4, and $\|g * K_\varepsilon - g\|_p \to 0$ ($\varepsilon \to 0$) can be obtained by Theorem 6.2.1.

Thus, the density can be deduced by $\|g * K_\varepsilon - f\|_p \le \|g * K_\varepsilon - g\|_p + \|h\|_p \le \|g * K_\varepsilon - g\|_p + \eta$. $\qquad\square$

Next, we discuss the identity approximation in the sense of pointwise and almost everywhere. In our discussion, we use the properties of Lebesgue points.

Lemma 6.2.4. *Let* $1 \le p < \infty$, $f \in L^p(\mathbb{R}^n)$. *If* x *is an* L^p *point of* f, *then* x *is an* L *point of* f *and* $HL(f)(x) < \infty$.

Proof. Let $0 < r < \infty$. By Hölder's inequality, we have

$$r^{-n} \int_{Q(x,r)} |f(y) - f(x)| \, dy \le \left(r^{-n} \int_{Q(x,r)} |f(y) - f(x)|^p \, dy \right)^{\frac{1}{p}},$$

which means that x is an L point of f, and there exists a positive number δ such that for any $0 < r \le \delta$,

$$r^{-n} \int_{Q(x,r)} |f(y) - f(x)| \, dy \le \left(r^{-n} \int_{Q(x,r)} |f(y) - f(x)|^p \, dy \right)^{\frac{1}{p}} < 1.$$

On the other hand, when $r > \delta$,

$$r^{-n} \int_{Q(x,r)} |f(y) - f(x)| \, dy \le \left(\delta^{-n} \int_{Q(x,r)} |f(y)|^p dy \right)^{\frac{1}{p}} + |f(x)|$$

$$\le \delta^{-\frac{n}{p}} \|f\|_p + |f(x)| < \infty,$$

which implies that

$$HL(f)(x) \leq |f(x)| + \sup_{r>0} r^{-n} \int_{Q(x,r)} |f(y) - f(x)|\, dy < \infty.$$

$\square$

Theorem 6.2.5. *Let* $K \in L(\mathbb{R}^n)$ *and* $\int_{\mathbb{R}^n} K(x)dx = 1$. *Define*

$$\varphi(r) = \sup\{|K(x)| : x \in \mathbb{R}^n,\ |x| \geq r\},\ r > 0.$$

If $\int_{\mathbb{R}^n} \varphi(|x|)\, dx < \infty$, *then for any* $f \in L^p(\mathbb{R}^n)$, $1 \leq p < \infty$, *we have*

$$\lim_{\varepsilon \to 0} (f * K_\varepsilon)(x) = f(x)$$

at the L^p *point* x *of* f.

Proof. Define

$$E = \{(y, t) \in \mathbb{R}^n \times (0, \infty) : \varphi(|y|) < t\},$$
$$E(t) = \{y \in \mathbb{R}^n : \varphi(|y|) > t\}, \quad t > 0,$$

then we have

$$\int_{\mathbb{R}^n} \varphi(|y|)\, dy = \int_0^\infty m(E(t))\, dt < \infty.$$

It can be seen that $E(t)$ is a ball with a finite radius centered at the origin. Let its radius be $r(t)$, and note $E(t) = B(r(t))$, then we have

$$|f * K_\varepsilon(x) - f(x)| = \left| \int_{\mathbb{R}^n} (f(x - \varepsilon y) - f(x))K(y)dy \right|$$

$$\leq \int_{\mathbb{R}^n} |f(x - \varepsilon y) - f(x)|\, \varphi(|y|)\, dy$$

$$= \int_{\mathbb{R}^n} |f(x - \varepsilon y) - f(x)| \int_0^\infty \chi_E(y, t)\, dt dy$$

$$= \int_0^\infty \frac{m(E(t))}{m(B(r(t)))} \int_{B(r(t))} |f(x - \varepsilon y) - f(x)|\, dy dt.$$

Write

$$g(x; \varepsilon, t) = \frac{1}{m(B(\varepsilon r(t)))} \int_{B(r(t))} |f(x - \varepsilon y) - f(x)|\, dy$$

$$= \frac{1}{m(B(\varepsilon r(t)))} \int_{B(\varepsilon r(t))} |f(x - y) - f(x)|\, dy.$$

According to Lemma 6.2.4, for any $\varepsilon > 0$, $t > 0$,

$$g(x; \varepsilon, t) \le C_n(HL(f)(x) + |f(x)|) < \infty, \quad \lim_{\varepsilon \to 0} g(x; \varepsilon, t) = 0.$$

Thus, by the Lebesgue dominated convergence theorem, we can get

$$\lim_{\varepsilon \to 0} |f * K_\varepsilon(x) - f(x)| = \int_0^\infty \lim_{\varepsilon \to 0} g(x; \varepsilon, t) m(E(t)) dt = 0.$$

$\square$

6.3 Poisson Integral and Further Application of HL

Poisson integral plays an important role in harmonic analysis. It's closely related to the theory of harmonic functions.

Definition 6.3.1 (Poisson kernel). Write

$$C_n = \pi^{-\frac{n+1}{2}} \Gamma\left(\frac{n+1}{2}\right),$$

$$P(x) = \frac{C_n}{(1 + |x|^2)^{\frac{n+1}{2}}}, \quad x \in \mathbb{R}^n.$$

P is called the n-dimensional Poisson kernel. For $y > 0$, let $P_y(x) = y^{-n} P(y^{-1} x)$. The convolution of f with P_y

$$(f * P_y)(x) = \int_{\mathbb{R}^n} f(x - t) P_y(t) \, dt, \ y > 0,$$

is called the Poisson integral of f.

It's easy to show that $P \in L(\mathbb{R}^n)$ and $\int_{\mathbb{R}^n} P(x) dx = 1$. Thus, by Theorem 6.2.1, we can obtain the following corollary.

Corollary 6.3.1. *If $f \in L^p(\mathbb{R}^n)$, $1 \le p < \infty$, then*

$$\lim_{y \to 0} \| f * P_y - f \|_p = 0.$$

The following corollary can be deduced from Theorem 6.2.2 directly.

Corollary 6.3.2. *If $f \in C_c(\mathbb{R}^n)$, then*

$$\lim_{y \to 0} (f * P_y)(x) = f(x)$$

holds on $\mathbb{R}^n$ uniformly.

It is obvious that the kernel P also satisfies the conditions of Theorem 6.2.5, so the theorem leads to the conclusion that Poisson integral converges almost everywhere. More precisely, we have the following corollary.

Corollary 6.3.3. *If $f \in L^p(\mathbb{R}^n)$, $1 \leq p < \infty$, then*

$$\lim_{y \to 0} (f * P_y)(x) = f(x)$$

holds for every L^p point x of f, and thus, the above equality holds for almost everywhere.

We know that in the boundary-value theory of analytic functions (or harmonic functions), nontangential limit is an important way. For example, the Poisson integral of an integrable function on a unit circle is analytic (harmonic) inside the circle and has nontangential boundary values almost everywhere. Without restriction to the limiting method, it may have no boundary values anywhere. In the upper half space, the situation is similar. In the following, we use the method of maximal functions to establish theorems about nontangential boundary values of Poisson integrals.

We write $\mathbb{R}_+^{n+1}$ as the upper half space of $(n+1)$-dimensional Euclidean space, that is,

$$\mathbb{R}_+^{n+1} = \{(x, y) : x \in \mathbb{R}^n,\ y > 0\}.$$

The region in the upper half space

$$\Gamma_\alpha(x_0) = \{(x, y) \in \mathbb{R}_+^{n+1} : |x - x_0| < \alpha y\},\ \alpha > 0,$$

is called a cone with x_0 as its vertex.

Definition 6.3.2. If for every α,

$$\lim_{\substack{(x,y) \to (x_0,0) \\ (x,y) \to \Gamma_\alpha(x_0)}} (f * P_y)(x) = f(x_0),$$

then we say that Poisson integral $(f * P_y)(x)$ converges nontangentially to $f(x_0)$.

Theorem 6.3.4. *If $f \in L^p(\mathbb{R}^n)$, $1 \leq p \leq \infty$, then for any $\alpha > 0$,*

$$\sup_{(x,y) \to \Gamma_\alpha(x_0)} \left| (f * P_y)(x) \right| \leq CHL(f)(x_0),$$

where C is independent of f and x_0, but dependent on n and α.

Proof. Let $(x, y) \in \Gamma_\alpha(x_0)$, then

$$|(f * P_y)(x)| = C_n \left| \int_{\mathbb{R}^n} \frac{y}{(|x - t|^2 + y^2)^{\frac{n+1}{2}}} f(t) dt \right|$$

$$\leq \frac{C_n}{y^n} \int_{|t-x_0|>2\alpha y} |f(t)| dt$$

$$+ C_n \int_{|t-x_0|>2\alpha y} \frac{y|f(t)|}{(|x - t|^2 + y^2)^{\frac{n+1}{2}}} dt$$

$$\leq CHL(f)(x_0) + C_n \int_{|t-x_0|>2\alpha y} \frac{y|f(t)|}{(|x - t|^2 + y^2)^{\frac{n+1}{2}}} dt.$$

Note that

$$\int_{|t-x_0|>2\alpha y} \frac{y|f(t)|}{(|x - t|^2 + y^2)^{\frac{n+1}{2}}} dt$$

$$= \sum_{k=1}^{\infty} \int_{2^k \alpha y < |t-x_0| \leq 2^{k+1}\alpha y} \frac{y|f(t)|}{(|x - t|^2 + y^2)^{\frac{n+1}{2}}} dt$$

$$\leq \sum_{k=1}^{\infty} (2^{k-1}\alpha y)^{-n-1} \int_{|t-x_0| \leq 2^{k+1}\alpha y} y|f(t)| dt$$

$$\leq 4^{n+1}\alpha^{-1} \sum_{k=1}^{\infty} 2^{-k}(2^{k+1}\alpha y)^{-n} \int_{|t-x_0| \leq 2^{k+1}\alpha y} |f(t)| dt$$

$$\leq CHL(f)(x_0). \qquad \square$$

Theorem 6.3.5. *If* $f \in L^p(\mathbb{R}^n)$, $1 \leq p < \infty$, *then for a.e.* $x_0 \in \mathbb{R}^n$,

$$\lim_{\substack{(x,y)\to(x_0,0) \\ (x,y)\to\Gamma_\alpha(x_0)}} (f * P_y)(x) = f(x_0)$$

holds for any $\alpha > 0$, *that is,* $f * P_y$ *converges nontangentially to* f *almost everywhere.*

Proof. We first show that if $f \in C_c^\infty(\mathbb{R}^n)$, then the conclusion of the theorem holds. In fact, when $f \in C_c^\infty(\mathbb{R}^n)$, for any $\varepsilon > 0$, there exists $\delta > 0$ such that

$$|f(u) - f(x_0)| < \varepsilon,$$

where

$$u \in B(x_0, \delta) = \{T : |x_0 - T| < \delta\}.$$

Then

$$|(f * P_y)(x) - f(x_0)|$$

$$= \left| \int_{\mathbb{R}^n} [f(x - t) - f(x_0)] P_y(t) dt \right|$$

$$\leq \int_{|t| < \frac{\delta}{2}} |f(x - t) - f(x_0)| P_y(t) dt$$

$$+ \int_{|t| \geq \frac{\delta}{2}} |f(x - t) - f(x_0)| P_y(t) dt.$$

When $|x - x_0| < \frac{\delta}{2}$ and $|t| < \frac{\delta}{2}$, we have $x - t \in B(x_0, \delta)$. Thus, if $|x - x_0| < \frac{\delta}{2}$,

$$\int_{|t| < \frac{\delta}{2}} |f(x - t) - f(x_0)| P_y(t) dt < \varepsilon.$$

Since $f \in C_c^\infty(\mathbb{R}^n)$,

$$\int_{|t| \geq \frac{\delta}{2}} |f(x - t) - f(x_0)| P_y(t) dt$$

$$\leq 2\|f\|_c \int_{|t| \geq \frac{\delta}{2y}} P(t) dt \to 0, \quad y \to 0.$$

Thus, for any $x_0 \in \mathbb{R}^n$ and $f \in C_c^\infty(\mathbb{R}^n)$, we have

$$\lim_{\substack{(x,y) \to (x_0, 0) \\ (x,y) \to \Gamma_\alpha(x_0)}} |(f * P_y)(x) - f(x_0)| = 0.$$

Now, suppose that $f \in L^p(\mathbb{R}^n)$, $1 \leq p < \infty$. By Corollary 6.2.3, there exists $g_k \in C_c^\infty(\mathbb{R}^n)$ such that

$$\lim_{k \to \infty} \|f - g_k\|_p = 0.$$

Then combining

$$|(f * P_y)(x) - f(x_0)| \leq |(f - g_k) * P_y(x)| + |(g_k * P_y)(x) - g_k(x_0)|$$

$$+ |g_k(x_0) - f(x_0)|$$

with Theorem 6.3.4, we obtain

$$\lim_{\substack{(x,y)\to(x_0,0)\\(x,y)\to\Gamma_\alpha(x_0)}} \left|(f * P_y)(x) - f(x_0)\right|$$

$$\leq \sup_{(x,y)\to\Gamma_\alpha(x_0)} \left|(f - g_k) * P_y(x)\right| + \left|g_k(x_0) - f(x_0)\right|$$

$$\leq CHL(f - g_k)(x_0) + \left|g_k(x_0) - f(x_0)\right|.$$

For any $\lambda > 0$, define

$$E(\lambda) = \{x_0 \in \mathbb{R}^n : \overline{\lim_y} |(f * P_y)(x) - f(x_0)| > \lambda\},$$

then

$$m(E(\lambda)) \leq m\left(\left\{x_0 \in \mathbb{R}^n : CHL(f - g_k)(x_0) > \frac{\lambda}{2}\right\}\right)$$

$$+ m\left(\left\{x_0 \in \mathbb{R}^n : |g_k(x_0) - f(x_0)| > \frac{\lambda}{2}\right\}\right)$$

$$\leq m\left(\left\{x_0 \in \mathbb{R}^n : HL(f - g_k)(x_0) > \frac{\lambda}{2C}\right\}\right)$$

$$+ \left(\frac{2}{\lambda}\right)^p \|f - g_k\|_p^p.$$

According to the weak type (p, p) property of HL, it is easy to prove

$$m(E(\lambda)) \leq C\lambda^{-p}\|f - g_k\|_p^p.$$

Note that when $k \to \infty$, $\|f - g_k\|_p \to \infty$, which implies that

$$m(E(\lambda)) = 0, \quad \lambda > 0.$$

Obviously, when $\alpha > \beta > 0$, we have $E(\alpha) \subset E(\beta)$, so

$$m\left(\bigcup_{\alpha>0} E(\alpha)\right) = m\left(\bigcup_{k=1}^{\infty} E\left(\frac{1}{k}\right)\right) = 0.$$

$\square$

Exercise 6

1. Let $0 < \alpha < \frac{n-1}{2}$, $n \geq 2$. Define T as follows:

$$T(f)(x) = \int_{\{y \in \mathbb{R}^n : |x-y| < 1\}} |x - y|^{-\left(\frac{n+1}{2} + \alpha\right)} f(y) dy.$$

   Prove that T is of type (p, p), $1 < p < \infty$.
2. Let $1 \leq p \leq \infty$, $f \in L^p(\mathbb{R}^n)$, $g \in L^{p'}(\mathbb{R}^n)$ with $\frac{1}{p} + \frac{1}{p'} = 1$. Prove that $f * g$ is bounded and continuous on $\mathbb{R}^n$.
3. Try to construct a function $f \in C_c^\infty(\mathbb{R}^n)$ such that f takes the maximum value at origin.
4. Let G_1, G_2 be bounded open subsets of $\mathbb{R}^n$, and $\overline{G}_1 \subset G_2$. Try to construct a function $f \in C_c^\infty(\mathbb{R}^n)$ such that

$$f(x) = \begin{cases} 1, & x \in G_1, \\ 0, & x \in \complement G_2. \end{cases}$$

5. Let $K \in L(\mathbb{R}^n)$ and $\int_{\mathbb{R}^n} K(x) dx = 1$. If $f \in C_c^m(\mathbb{R}^n)$, then $D^\alpha(f * K_\varepsilon)(x)$ converges uniformly to $D^\alpha f(x)$ ($\varepsilon \to 0$) on $\mathbb{R}^n$, where $|\alpha| \leq m$.
6. Let $\{T_\varepsilon\}_{(\varepsilon > 0)} : L^1(\mathbb{R}^n) \to \mathscr{F}$ (a class of measurable functions) be a family of linear operators, define T^*:

$$T^*(f)(x) = \sup_{\varepsilon > 0} |T_\varepsilon(f)(x)|.$$

   Prove that if T^* is of weak type $(1, 1)$ and

$$\lim_{\varepsilon \to 0} T_\varepsilon(g)(x) = g(x) \quad a.e.$$

   holds for any $g \in C(\mathbb{R}^n) \cap L^1(\mathbb{R}^n)$, then the equality holds for any $f \in L^1(\mathbb{R}^n)$.
7. Prove that if $f \in L^p(\mathbb{R}^n)$, $1 \leq p < \infty$, then

$$\int_{\mathbb{R}^n} |f(x - t) - f(x)|^p dx \to 0, \quad |t| \to 0.$$

Chapter 7

The Fourier Transform

7.1 Fourier Transform on $L^1(\mathbb{R}^n)$

Let $x = (x_1, \ldots, x_n)$, $y = (y_1, \ldots, y_n) \in \mathbb{R}^n$. Write

$$x \cdot y = x_1 y_1 + \cdots + x_n y_n, \quad |x| = \sqrt{x \cdot x}.$$

Definition 7.1.1. Let $f \in L^1(\mathbb{R}^n)$, then

$$\hat{f}(x) = \int_{\mathbb{R}^n} f(y) e^{-2\pi i x \cdot y} dy$$

is called the Fourier transform of f, denoted as $\hat{f} = \mathscr{F}(f)$.

Theorem 7.1.1. If $f \in L^1(\mathbb{R}^n)$, then

(1) $\hat{f} \in L^\infty(\mathbb{R}^n)$ and $\|\hat{f}\|_\infty \leq \|f\|_1$;
(2) $\hat{f}$ is continuous uniformly on $\mathbb{R}^n$;
(3) $\lim_{|x| \to \infty} \hat{f}(x) = 0$.

Proof. (1) By Definition 7.1.1 and

$$\hat{f}(x+h) - \hat{f}(x) = \int_{\mathbb{R}^n} f(y) e^{-2\pi i x \cdot y} (e^{-2\pi i h \cdot y} - 1) dy,$$

we can obtain

$$|\hat{f}(x+h) - \hat{f}(x)| \leq \int_{\mathbb{R}^n} |f(y)| \, |2 \sin(\pi y \cdot h)| dy.$$

The right-hand side of above inequality is independent of x, and when $h \to 0$, by the Lebesgue dominated convergence theorem, it is an infinitely small magnitude, so (2) holds. Since the family of simple functions that has the form

$$f(x) = \sum_{k=1}^{m} c_k \chi_{Q_k}(x)$$

is dense in $L^1(\mathbb{R}^n)$, where Q_k is a cube, then to prove (3), we only need to prove that (3) holds for the characteristic function χ_Q of any cube Q. Thus, we can finish our proof by calculating the repeated integral of $\widehat{\chi_Q}(x)$. □

Theorem 7.1.2 (Multiplication formula). *Let* $f, g \in L^1(\mathbb{R}^n)$, *then*

$$\int_{\mathbb{R}^n} f(x)\widehat{g}(x)\, dx = \int_{\mathbb{R}^n} \widehat{f}(x)g(x)dx.$$

Proof. By Fubini's theorem,

$$\int_{\mathbb{R}^n} f(x)\widehat{g}(x)\, dx = \int_{\mathbb{R}^n} f(x)\left(\int_{\mathbb{R}^n} g(y)e^{-2\pi i y\cdot x}\, dy\right) dx$$

$$= \int_{\mathbb{R}^n} \left(\int_{\mathbb{R}^n} f(x)e^{-2\pi i y\cdot x}\, dx\right) g(y)\, dy$$

$$= \int_{\mathbb{R}^n} \widehat{f}(y)g(y)\, dy.$$

□

Theorem 7.1.3 (Derivative formula). *Let* $f \in L^1(\mathbb{R}^n)$. *If* x_k $f(x) \in L^1(\mathbb{R}^n)$ $(k \in \{1, \ldots, n\})$, *then*

$$\frac{\partial \widehat{f}(x)}{\partial x_k} = \mathscr{F}(-2\pi i y_k f(y))(x).$$

Proof. Write

$$\Delta_k = (0, \ldots, 0, \delta, 0, \ldots, 0), \quad \delta > 0, \quad \delta \text{ is the } k\text{th element.}$$

Then

$$\frac{\widehat{f}(x + \Delta_k) - \widehat{f}(x)}{\delta} = \int_{\mathbb{R}^n} f(y)\frac{1}{\delta}[e^{-2\pi i y\cdot(x+\Delta_k)} - e^{-2\pi i y\cdot x}]\, dy$$

$$= \int_{\mathbb{R}^n} f(y)e^{-2\pi i y\cdot x}\frac{e^{-2\pi i y_k \delta} - 1}{\delta}\, dy.$$

According to

$$\left|\frac{1}{\delta}(e^{-2\pi i y_k \delta} - 1)\right| \leq 2\pi |y_k|$$

and $y_k f(y) \in L^1(\mathbb{R}^n)$, together with the Lebesgue dominated convergence theorem, we have

$$\frac{\partial \widehat{f}(x)}{\partial x_k} = \lim_{\delta \to 0} \frac{1}{\delta}\left[\widehat{f}(x + \Delta_k) - \widehat{f}(x)\right]$$

$$= \int_{\mathbb{R}^n} f(y)e^{-2\pi i y \cdot x} \lim_{\delta \to 0} \frac{e^{-2\pi i y_k \delta} - 1}{\delta} \, dy$$

$$= \int_{\mathbb{R}^n} (-2\pi i y_k f(y))e^{-2\pi i y \cdot x} \, dy$$

$$= \mathscr{F}(-2\pi i y_k f(y))(x). \qquad \square$$

Theorem 7.1.4 (Derivative formula). *Let* $f \in L^1(\mathbb{R}^n)$. *If* f *is absolutely continuous about the kth variety on* $\mathbb{R}$ $(k \in \{1, \ldots, n\})$, *then*

$$\mathscr{F}\left(\frac{\partial f(y)}{\partial y_k}\right)(x) = 2\pi i x_k \widehat{f}(x).$$

Proof. We have

$$\mathscr{F}\left(\frac{\partial f(y)}{\partial y_k}\right)(x) = \int_{\mathbb{R}^n} \frac{\partial f(y)}{\partial y_k} e^{-2\pi i y \cdot x} \, dy$$

$$= \int_{\mathbb{R}^{n-1}} e^{-2\pi i \sum_{m \neq k} y_m x_m}$$

$$\times \left(\int_{\mathbb{R}} \frac{\partial f(y)}{\partial y_k} e^{-2\pi i y_k x_k} \, dy_k\right) \prod_{m \neq k} dy_m.$$

Integrating by parts yields that

$$\int_{\mathbb{R}} \frac{\partial f(y)}{\partial y_k} e^{-2\pi i y_k x_k} \, dy_k = 2\pi i x_k \int_{\mathbb{R}} f(y)e^{-2\pi i y_k x_k} \, dy_k.$$

Then

$$\mathscr{F}\left(\frac{\partial f(y)}{\partial y_k}\right)(x) = 2\pi i x_k \widehat{f}(x). \qquad \square$$

Remark. Theorem 7.1.3 indicates the formula of taking the derivative of the Fourier transform. Theorem 7.1.4 shows the formula of taking the Fourier transform of the derivative.

Theorem 7.1.5 (Fourier transform of convolution). *Let* $f, g \in L^1(\mathbb{R}^n)$, *then*

$$\mathscr{F}(f * g)(x) = \mathscr{F}(f) \cdot \mathscr{F}(g).$$

Proof. By Fubini's theorem,

$$\mathscr{F}(f * g)(x) = \int_{\mathbb{R}^n} \left(\int_{\mathbb{R}^n} f(u - y) g(y) \, dy \right) e^{-2\pi i x \cdot u} \, du$$

$$= \int_{\mathbb{R}^n} g(y) \left(\int_{\mathbb{R}^n} f(u - y) e^{-2\pi i x \cdot u} \, du \right) dy$$

$$= \int_{\mathbb{R}^n} g(y) \left(\int_{\mathbb{R}^n} f(v) e^{-2\pi i x \cdot (v+y)} \, dv \right) dy$$

$$= \left(\int_{\mathbb{R}^n} f(v) e^{-2\pi i x \cdot v} \, dv \right) \left(\int_{\mathbb{R}^n} g(y) e^{-2\pi i x \cdot y} \, dy \right).$$

□

Remark. The importance of Theorem 7.1.5 is that Fourier transform changes the convolution operation into the product operation.

Let τ_h be a translation transform, which means that $\tau_h f(x) = f(x - h)$.

Theorem 7.1.6. *Let* $f \in L^1(\mathbb{R}^n)$, *then*

(i) $(\tau_h f)^\wedge(x) = e^{-2\pi i x \cdot h} \widehat{f}(x)$,
(ii) $(\tau_h \widehat{f})(x) = \left(e^{2\pi i t \cdot h} f(t) \right)^\wedge(x)$.

Proof. The theorem can be obtained by the definitions of Fourier transform and translation transform directly. □

Remark. The equality (i) indicates the formula of taking the Fourier transform of the translation transform. The equality (ii) indicates the formula of taking the translation transform of the Fourier transform.

Let δ_a be a dilation transform $(a > 0)$, which means that $\delta_a f(x) = f(ax)$. The definitions of Fourier transform and dilation transform yield the following theorem directly.

Theorem 7.1.7. $(\delta_a f)^\wedge(x) = a^{-n}\widehat{f}(\frac{x}{a})$.

Example 7.1.1. Let $\Phi(x) = e^{-2\pi|x|}$, then $\widehat{\Phi}(x) = P(x)$, where $P(x)$ is n-dimensional varieties Poisson kernel.

Proof. Note that

$$e^{-t} = \frac{1}{\sqrt{\pi}} \int_0^\infty \frac{e^{-u}}{\sqrt{u}} e^{-\frac{t^2}{4u}} \, du, \ t \in \mathbb{R}. \tag{$*$}$$

The identity $(*)$ can be deduced by the circulatory integral of the function

$$\varphi(z) = \frac{1}{1+z^2} e^{itz}$$

on the upper half circle of the complex plane with the origin as the center and R $(R > 1)$ as the radius and residue theorem. Then, by $(*)$,

$$\widehat{\Phi}(x) = \int_{\mathbb{R}^n} e^{-2\pi|y|} e^{-2\pi i x \cdot y} \, dy$$

$$= \int_{\mathbb{R}^n} \left(\frac{1}{\sqrt{\pi}} \int_0^\infty \frac{e^{-u}}{\sqrt{u}} e^{-\pi^2 \frac{|y|^2}{u}} \, du \right) e^{-2\pi i x \cdot y} \, dy$$

$$= \frac{1}{\sqrt{\pi}} \int_0^\infty \frac{e^{-u}}{\sqrt{u}} \left(\int_{\mathbb{R}^n} e^{-\pi^2 \frac{|y|^2}{u}} e^{-2\pi i x \cdot y} \, dy \right) du$$

$$= \frac{1}{\sqrt{\pi}} \int_0^\infty \frac{e^{-u}}{\sqrt{u}} \left(\prod_{j=1}^n \int_{-\infty}^\infty e^{-\pi^2 \frac{y_j^2}{u} - 2\pi i x_j y_j} \, dy_j \right) du$$

$$= \frac{1}{\sqrt{\pi}} \int_0^\infty \frac{e^{-u}}{\sqrt{u}} \left(\prod_{j=1}^n \int_{-\infty}^\infty e^{-(\pi \frac{y_j}{\sqrt{u}} + \sqrt{u} i x_j)^2 - u x_j^2} \, dy_j \right) du$$

$$= \frac{1}{\sqrt{\pi}} \int_0^\infty \frac{e^{-u}}{\sqrt{u}} \left(\prod_{j=1}^n \left(e^{-u x_j^2} \cdot \frac{\sqrt{u}}{\pi} \sqrt{\pi} \right) \right) du$$

$$= \pi^{-\frac{n+1}{2}} \int_0^\infty u^{\frac{n+1}{2}} e^{-u(1+|x|^2)} \, du$$

$$= \frac{1}{\pi^{\frac{n+1}{2}}} \frac{1}{(1+|x|^2)^{\frac{n+1}{2}}} \int_0^\infty s^{\frac{n-1}{2}} e^{-s} \, ds$$

$$= P(x). \qquad \qquad \square$$

Theorem 7.1.8 (Fourier inversion formula). *Let $f \in L^1(\mathbb{R}^n)$. If $\widehat{f} \in L^1(\mathbb{R}^n)$, then*

$$f(x) = \int_{\mathbb{R}^n} \widehat{f}(y) e^{2\pi i x \cdot y} \, dy, \quad a.e.$$

Furthermore, if f is a continuous function, the above equality holds everywhere.

Proof. Write

$$Q_\varepsilon(f)(x) = \int_{\mathbb{R}^n} \widehat{f}(y) e^{2\pi i x \cdot y} e^{-2\pi \varepsilon |y|} \, dy, \quad \varepsilon > 0.$$

By the Lebesgue dominated convergence theorem, we can get

$$\lim_{\varepsilon \to 0} Q_\varepsilon(f)(x) = \int_{\mathbb{R}^n} \widehat{f}(y) e^{2\pi i x \cdot y} \, dy.$$

On the other hand, by Theorems 7.1.2, 7.1.6 and 7.1.7, we have

$$Q_\varepsilon(f)(x) = \int_{\mathbb{R}^n} f(y) \left(e^{2\pi i x \cdot t} e^{-2\pi \varepsilon |t|} \right)^\wedge (y) \, dy$$

$$= \int_{\mathbb{R}^n} f(y) \tau_x \widehat{e^{-2\pi \varepsilon |\cdot|}}(y) \, dy$$

$$= \int_{\mathbb{R}^n} f(y) \tau_x \delta_\varepsilon \widehat{e^{-2\pi |\cdot|}}(y) \, dy$$

$$= \int_{\mathbb{R}^n} f(y) \tau_x \varepsilon^{-n} \widehat{e^{-2\pi |\cdot|}} \left(\frac{y}{\varepsilon} \right) \, dy$$

$$= \int_{\mathbb{R}^n} f(y) P_\varepsilon(x - y) \, dy = (f * P_\varepsilon)(y),$$

where P is Poisson kernel. Then, by Corollary 6.3.3,

$$\lim_{\varepsilon \to 0} Q_\varepsilon(f)(x) = f(x), \quad a.e.$$

Thus,

$$f(x) = \int_{\mathbb{R}^n} \widehat{f}(y) e^{2\pi i x \cdot y} \, dy, \quad \text{a.e.}$$

Obviously, if f is a continuous function, the above equality holds everywhere. □

7.2 Fourier Transform on $L^2(\mathbb{R}^n)$

As we all know, the nest between $L^1(\mathbb{R}^n)$ and $L^2(\mathbb{R}^n)$ is invalid. But $C_c^\infty(\mathbb{R}^n)$ is dense in $L^p(\mathbb{R}^n)$, $1 \le p \le \infty$. Then $L^1(\mathbb{R}^n) \cap L^2(\mathbb{R}^n)$ is dense in $L^2(\mathbb{R}^n)$. According to this fact, the definition of fourier transform can be extended to $L^2(\mathbb{R}^n)$.

Theorem 7.2.1 (Plancherel's theorem). *If* $f \in L^1(\mathbb{R}^n) \cap L^2(\mathbb{R}^n)$, *then* $\widehat{f} \in L^2(\mathbb{R}^n)$, *and*

$$\|\widehat{f}\|_2 = \|f\|_2.$$

Proof. Let $g(x) = f(-x)$ and $\bar{f}(x)$ be the complex conjugate of $f(x)$. Define $h = g * \bar{f}$, by Hölder's inequality, we can get

$$|h(x) - h(x')| = \left| \int_{\mathbb{R}^n} [f(y - x) - f(y - x')] \bar{f}(y) \, dy \right|$$

$$\le \|f(\cdot - x) - f(\cdot - x')\|_2 \|f\|_2.$$

Then, $h \in C(\mathbb{R}^n) \cap L^1(\mathbb{R}^n)$. Theorem 7.1.5 yields that

$$\mathscr{F}(h) = \mathscr{F}(g)\mathscr{F}(\bar{f}).$$

But by $\widehat{g}(x) = \widehat{f}(-x)$, and

$$\mathscr{F}(\bar{f})(x) = \int_{\mathbb{R}^n} \overline{f(y)} e^{-2\pi i x \cdot y} dy = \overline{\widehat{f}(-x)},$$

$\widehat{h}(x) = |\widehat{f}(-x)|^2 \ge 0$ and $\widehat{h} \in L^1(\mathbb{R}^n)$. Note that $h \in C(\mathbb{R}^n) \cap L^1(\mathbb{R}^n)$, and by Fourier reversion formula,

$$h(0) = \int_{\mathbb{R}^n} \widehat{h}(x) \, dx,$$

where $h(0) = \|f\|_2^2$ and $\int_{\mathbb{R}^n} \widehat{h}(x) \, dx = \|\widehat{f}\|_2^2$. □

By Theorem 7.2.1, Fourier transform $\mathcal{F} : L^1 \cap L^2 \to L^2$ is a bounded linear operator with norm 1. Then $\mathcal{F}$ can be uniquely extended from the dense subspace $L^1 \cap L^2$ to L^2 with the same norm. The operator, obtained by the extension, is defined as the Fourier transform on L^2 and can be written by $\mathcal{F}$ as well. By the Fourier reversion formula, $\mathcal{F}$ is injective mapping from L^2 to itself (different elements must have different images). Now, let us prove that $\mathcal{F}$ is a surjective mapping. For this, we prove the following lemma at first.

Lemma 7.2.2. *Let*

$$A = (a_1, \ldots, a_n) \in \mathbb{R}^n, \quad B = (b_1, \ldots, b_n) \in \mathbb{R}^n,$$

and $a_j < b_j$, $1 \le j \le n$. Write

$$\varphi_{a,b}(t) = \frac{1}{2\pi i t}(e^{2\pi i b t} - e^{2\pi i a t}), \quad -\infty < a < b < \infty, \ t \in \mathbb{R},$$

and

$$\varphi_{A,B}(x) = \prod_{j=1}^{n} \varphi_{a_j,b_j}(x_j), \quad x \in \mathbb{R}^n,$$

then $\varphi_{A,B} \in L^2(\mathbb{R}^n)$, and $\mathcal{F}(\varphi_{A,B}) = \chi_{A,B}$, where

$$\chi_{A,B}(x) = \prod_{i=1}^{n} \chi_{a_j,b_j}(x_j)$$

and

$$\chi_{a_j,b_j}(x_j) = \begin{cases} 1, & a_j < x_j < b_j, \\ \dfrac{1}{2}, & x_j = a_j \ or \ x_j = b_j, \\ 0, & x_j \notin [a_j, b_j]. \end{cases}$$

Proof. We consider only the case $n = 1$. According to $\varphi_{a,b}(t) \in C(\mathbb{R})$, and when $t \to \infty$,

$$\varphi_{a,b}(t) = O\left(\frac{1}{t}\right),$$

then $\varphi_{a,b} \in L^2(\mathbb{R}^n)$. Suppose $m \in \mathbb{N}$, and write

$$\varphi_{a,b}^m(t) = \begin{cases} \varphi_{a,b}(t), & |t| < m, \\ 0, & |t| \ge m. \end{cases}$$

Then

$$\varphi_{a,b}^m \in L^1(\mathbb{R}) \cap L^2(\mathbb{R})$$

and

$$\lim_{m \to \infty} \varphi_{a,b}^m \overset{L^2}{=\!=\!=} \varphi_{a,b}.$$

By the definition of Fourier transform on L^2,

$$\mathscr{F}(\varphi_{a,b}) \overset{L^2}{=\!=\!=} \lim_{m \to \infty} \mathscr{F}(\varphi_{a,b}^m).$$

The following equality holds pointwise:

$$\lim_{m \to \infty} \mathscr{F}(\varphi_{a,b}^m)(s) = \lim_{m \to \infty} \int_{-m}^m \frac{e^{2\pi i b t} - e^{2\pi i a t}}{2\pi i t} e^{-2\pi i s t} \, dt$$

$$= \lim_{m \to \infty} \int_0^m \frac{\sin 2\pi(b-s)t - \sin 2\pi(a-s)t}{\pi t} \, dt$$

$$= \chi_{a,b}(s).$$

Therefore, $\mathscr{F}(\varphi_{a,b}) = \chi_{a,b}$. □

Theorem 7.2.3. *The mapping $\mathscr{F}$ is the surjective mapping from L^2 to itself.*

Proof. The linear manifold generated by the function $\chi_{A,B}$ in Lemma 7.2.2 is dense in L^2, and every $\chi_{A,B}$ is an image of $\mathscr{F}$. Then we only need to prove that the image of L^2 under the action of $\mathscr{F}$ is a closed set of L^2. However, the latter is obvious. In fact, $\mathscr{F}$ as a bounded operator on L^2 maintains the norm of 1. □

Thus, $\mathscr{F}$ is an isometry on L^2. We denote the inverse mapping of $\mathscr{F}$ by $\mathscr{F}^{-1}$, which is also a bounded linear operator with norm 1 on L^2.

Theorem 7.2.4 (Plancherel's theorem). *Let $f \in L^2(\mathbb{R}^n)$, then*

$$(\mathscr{F}^{-1}f)(x) = (\mathscr{F}f)(-x).$$

Proof. Suppose $f \in L^1(\mathbb{R}^n) \cap L^2(\mathbb{R}^n)$. Write

$$h(x) = (\mathscr{F}f)(-x) = \int_{\mathbb{R}^n} f(y) e^{2\pi i x \cdot y} \, dy.$$

Let $g \in L^1(\mathbb{R}^n) \cap L^2(\mathbb{R}^n)$, then

$$
\begin{aligned}
(g, h) &= \int_{\mathbb{R}^n} g(x)\overline{h(x)}\, dx \\
&= \int_{\mathbb{R}^n} g(x)\left[\int_{\mathbb{R}^n} \overline{f(y)}e^{-2\pi ix \cdot y}\, dy\right] dx \\
&= \int_{\mathbb{R}^n} \overline{f(y)}\widehat{g}(y)\, dy = (\widehat{g}, f).
\end{aligned}
$$

Let $h^* = \mathscr{F}^{-1}(f)$, then $f = \widehat{h^*}$. Thus,

$$(g, h) = (\widehat{g}, \widehat{h^*}).$$

Since $\mathscr{F}$ is an isometric isomorphism mapping on L^2, we have

$$(\widehat{g}, \widehat{h^*}) = (g, h^*).$$

From this,

$$(g, h) = (g, h^*)$$

holds for any $g \in L^1(\mathbb{R}^n) \cap L^2(\mathbb{R}^n)$. According to the fact that $L^1 \cap L^2$ is dense in L^2, the above equality holds for any $g \in L^2(\mathbb{R}^n)$. Then we deduce $h^* = h$, that is,

$$(\mathscr{F}^{-1}f)(x) = (\mathscr{F}f)(-x), \quad f \in L^1(\mathbb{R}^n) \cap L^2(\mathbb{R}^n).$$

Using again the density of $L^1 \cap L^2$ in L^2, we know that above equality holds for any $f \in L^2(\mathbb{R}^n)$. $\qquad \square$

Applying the definition of Fourier transform on $L^1(\mathbb{R}^n)$ and $L^2(\mathbb{R}^n)$, we can naturally extend the definition of Fourier transform to $L^p(\mathbb{R}^n)$, $1 < p < 2$. In fact, when $1 < p < 2$,

$$L^p(\mathbb{R}^n) \subset L^1(\mathbb{R}^n) + L^2(\mathbb{R}^n),$$

which means that for any $f \in L^p(\mathbb{R}^n)$, $1 < p < 2$, there exists a decomposition of f: $f = f_1 + f_2$, such that $f_1 \in L^1(\mathbb{R}^n)$ and $f_2 \in L^2(\mathbb{R}^n)$. Actually, we can choose

$$
\begin{aligned}
f_1(x) &= f(x)\chi_{\{|f(x)|\geq 1\}}(x), \\
f_2(x) &= f(x)\chi_{\{|f(x)|<1\}}(x).
\end{aligned}
$$

Thus, we define

$$\mathscr{F}f = \mathscr{F}f_1 + \mathscr{F}f_2.$$

By the definition, it's easy to extend Theorems 7.1.2 and 7.1.5 to the following two theorems, respectively.

Theorem 7.2.5. *Let* $f \in L^p(\mathbb{R}^n)$, $1 \leq p \leq 2$. *If* $g \in L^1(\mathbb{R}^n)$, *then*

$$\mathscr{F}(f * g)(x) = \mathscr{F}(f)(x) \cdot \mathscr{F}(g)(x).$$

Theorem 7.2.6. *Let* $f \in L^2(\mathbb{R}^n)$, $g \in L^2(\mathbb{R}^n)$, *then*

$$\int_{\mathbb{R}^n} f(x)\widehat{g}(x)\, dx = \int_{\mathbb{R}^n} \widehat{f}(x)g(x)\, dx.$$

Finally, we establish a very useful inequality about Fourier transform by the Riesz–Thörin theorem.

Theorem 7.2.7 (Hausdorff–Young inequality). *Let* $f \in L^p$ $(\mathbb{R}^n)$, $1 \leq p \leq 2$, *then* $\widehat{f} \in L^{p'}(\mathbb{R}^n)$ *and*

$$\|\widehat{f}\|_{p'} \leq \|f\|_p.$$

Proof. By Theorem 7.1.1 (1), $\mathscr{F}$ is of type $(1, \infty)$ and

$$\|\widehat{f}\|_\infty \leq \|f\|_1.$$

According to Theorem 7.2.1 and the definition of Fourier transform on L^2, $\mathscr{F}$ is of type $(2, 2)$ and

$$\|\widehat{f}\|_2 = \|f\|_2.$$

Choose

$$(p_0, q_0) = (1, \infty), \quad (p_1, q_1) = (2, 2), \quad (p_t, q_t) = (p, p'),$$

then the Riesz–Thörin theorem yields that

$$\|\widehat{f}\|_{p'} \leq \|f\|_p. \qquad \square$$

7.3   An Application of Fourier Integral

For $f \in L(\mathbb{R}^n)$, the integral form

$$\int_{\mathbb{R}^n} \widehat{f}(y)e^{2\pi i x \cdot y}\, dy \qquad (7.1)$$

is called the Fourier integral of f. From Fourier inversion formula (Theorem 7.1.8), we can see that when $f \in L^1(\mathbb{R}^n)$ and $\widehat{f} \in L^1(\mathbb{R}^n)$, the Fourier integral of f exists and

$$\int_{\mathbb{R}^n} \widehat{f}(y) e^{2\pi i x \cdot y} \, dy = f(x), \quad \text{a.e.}$$

However, in general case, $\widehat{f} \notin L^1(\mathbb{R}^n)$, and the Fourier integral of f is just a formal integral. In this case, how to express f through $\widehat{f}$ becomes a very meaningful question. Write a Lebesgue integral by multiplying a function $\Phi(\varepsilon x)$, which has some interesting properties and depends on the parameter $\varepsilon > 0$, into the integral of (7.1):

$$\Phi_\varepsilon(f)(x) = \int_{\mathbb{R}^n} \widehat{f}(y) e^{2\pi i x \cdot y} \Phi(\varepsilon y) \, dy. \tag{7.2}$$

Then let $\varepsilon \to 0_+$, consider whether the limit is equal to f on a certain scale. This method is often called the sum of Fourier integral. It is closely related to the properties of Φ. Write $\widehat{\Phi} = \varphi$, we can rewrite (7.2) by the properties of Fourier transform as

$$\begin{aligned}
\Phi_\varepsilon(f)(x) &= \int_{\mathbb{R}^n} \widehat{f}(y) e^{2\pi i x \cdot y} \Phi(\varepsilon y) \, dy \\
&= \int_{\mathbb{R}^n} f(y) \mathscr{F}\left(e^{2\pi i x \cdot u} \Phi(\varepsilon u)\right)(y) \, dy \\
&= \int_{\mathbb{R}^n} f(y) \tau_x \mathscr{F}\left(\Phi(\varepsilon \cdot)\right)(y) \, dy \\
&= \int_{\mathbb{R}^n} f(y) \tau_x \mathscr{F}\left(\delta_\varepsilon \Phi(\cdot)\right)(y) \, dy \\
&= \int_{\mathbb{R}^n} f(y) \varepsilon^{-n} \varphi\left(\frac{y - x}{\varepsilon}\right) dy \\
&= (f * \varphi_\varepsilon)(x)
\end{aligned} \tag{7.3}$$

by Theorems 7.1.2, 7.1.6 and 7.1.7. Then the question whether the integral (7.2) converges (on a certain scale) to f is transformed into the question whether the sequence of convolution operators converges or the identity approximation operator. Therefore, according to Theorems 6.2.1 ($p = 1$) and 6.2.4 ($p = 1$), the following theorem is deduced.

Theorem 7.3.1. *Let* $\Phi \in L^1(\mathbb{R}^n)$, $\widehat{\Phi} = \varphi \in L^1(\mathbb{R}^n)$, *and*

$$\Phi(0) = \int_{\mathbb{R}^n} \varphi(x)\,dx = 1.$$

(a) *If* $f \in L^1(\mathbb{R}^n)$, *then*

$$\lim_{\varepsilon \to 0^+} \|\Phi_\varepsilon(f) - f\|_1 = 0.$$

(b) *If* $f \in L^1(\mathbb{R}^n)$, *and for some* $\lambda > 0$,

$$\varphi(x) = O((1 + |x|)^{-n-\lambda}),$$

then

$$\lim_{\varepsilon \to 0_+} \Phi_\varepsilon(f)(x) = f(x) \quad a.e.$$

If $\Phi \in L^1(\mathbb{R}^n) \cap L^2(\mathbb{R}^n)$, then Theorem 7.3.1 can be further generalized using Fourier transform theory on L^2. For this, let $f \in L^1(\mathbb{R}^n) + L^2(\mathbb{R}^n)$, there exists a decomposition of f:

$$f = f_1 + f_2,$$

where $f_1 \in L^1(\mathbb{R}^n)$, $f_2 \in L^2(\mathbb{R}^n)$. By (7.3), we can obtain

$$\Phi_\varepsilon(f_1)(x) = (f_1 * \varphi_\varepsilon)(x).$$

Then, by Plancherel's theorem, we know that $\varphi \in L^2(\mathbb{R}^n)$. Note that Theorems 7.1.6 and 7.1.7 hold for $f \in L^2(\mathbb{R}^n)$. Thus, using Theorems 7.1.6, 7.1.7 and 7.2.6 and by the way similar to the proof of equality (7.3), we can get

$$\Phi_\varepsilon(f_2)(x) = (f_2 * \varphi_\varepsilon)(x).$$

Therefore,

$$\Phi_\varepsilon(f)(x) = (f * \varphi_\varepsilon)(x).$$

Thus, we deduce the following theorem.

Theorem 7.3.2. *Let* $\Phi \in L^1(\mathbb{R}^n) \cap L^2(\mathbb{R}^n)$. *If*

$$\widehat{\Phi} = \varphi \in L^1(\mathbb{R}^n), \quad \Phi(0) = \int_{\mathbb{R}^n} \varphi(x)\,dx = 1,$$

and $f \in L^p(\mathbb{R}^n)$, $1 < p < 2$, *then*

$$\lim_{\varepsilon \to 0} \|\Phi_\varepsilon(f) - f\|_p = 0.$$

Furthermore, for some $\lambda > 0$,

$$\varphi(x) = O\left((1 + |x|)^{-n-\lambda}\right),$$

then

$$\lim_{\varepsilon \to 0} \Phi_\varepsilon(f)(x) = f(x), \quad a.e.$$

Exercise 7

1. Let $f(t) = \chi_{(a,b)}(t)$, $t \in \mathbb{R}$. Prove that $\hat{f} \notin L(\mathbb{R})$.
2. Let $f \in L^1(\mathbb{R}^n)$ and $\hat{f} \in L^1(\mathbb{R}^n)$. Show that

$$\mathscr{F}^{-1}(\mathscr{F}f(x)) = \mathscr{F}(\mathscr{F}^{-1}f(x)) = f(x). \quad \text{a.e.}$$

3. Let $f, g \in L^2(\mathbb{R}^n)$. If $\hat{f} = \hat{g}$, then

$$f = g. \quad \text{a.e.}$$

4. If $f, g \in L^2(\mathbb{R}^n)$, then

$$\mathscr{F}(f \cdot g) = \hat{f} * \hat{g}.$$

5. If $f, g \in L^2(\mathbb{R}^n)$, then

$$\mathscr{F}^{-1}(\hat{f} \cdot \hat{g}) = f * g.$$

6. Let $\Phi(x) = e^{-2\pi|x|}$, then Φ satisfies the conditions of Theorem 7.3.2.

Bibliography

[1] R. B. Ash, *Measure, Integration and Functional Analysis*, Academic Press, New York, 1972.

[2] D. L. Cohn, *Measure Theory*, Birkhäuser, Boston, 1980.

[3] G. B. Folland, *Real Analysis*, John Wiley & Sons, Inc., New York, 1984.

[4] M. de Guzmán, *Real Variable Method in Fourier Analysis*, North-Holland Publishing Company, Amsterdam-New York, 1981.

[5] E. Hewitt and K. R. Strömberg, *Real and Abstract Analysis*, Springer-Verlag, New York-Heidelberg, 1975.

[6] H. L. Royden, *Real Analysis*, 2nd edition, The MacMillan Co., New York, 1968.

[7] W. Rudin, *Real and Complex Analysis*, 2nd edition, McGraw-Hill, New York, 1974.

[8] I. E. Segal and R. A. Kunze, *Integrals and Operators*, Springer-Verlag, Berlin-New York, 1968.

[9] R. L. Wheeden and A. Zygmund, *Measure and Integral*, CRC Press, New York, 1977.

[10] 江泽涵, 拓扑学引论, 上海科学技术出版社, 1978.

[11] 方嘉琳, 点集拓扑学, 辽宁人民出版社, 1983.

[12] 钱佩玲, 柳藩, 实变函数论与泛函分析, 北京师范大学出版社, 1987.

[13] 孙永生, 泛函分析讲义, 北京师范大学出版社, 1986.

[14] 张禾瑞, 近世代数基础, 人民教育出版社, 1978.

[15] W. Fleming, *Functions of Several Variables*, Springer-Verlag, Springer New York, NY, 1977, p. 140.

Index